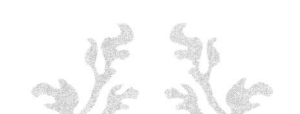

PEPTIDES MADE EASY

A Beginner's Guide to Understanding, Synthesizing, and Applying Peptides in Science and Wellness

HealthPoint Publications

TABLE OF CONTENTS

Chapter 1: Introduction

Peptides are vital molecules that play a central role in biology and chemistry. They serve as crucial messengers, regulators, and structural elements in highly complex physiological (and biochemical) processes found in living organisms. Those unique properties and extensive functions made them a target for studies in artificial biology, biotechnology, and pharmaceutical sciences. Interest in peptides has increased recently because of their versatility and potential for emerging applications, such as new drugs or healing therapies. The bankruptcy presents a considerable overview of peptides, ranging from their definition and introductory chemistry to their classifications, organic features, and structural homes. With this information, you will appreciate peptides' essential roles in each natural and artificial biological system.

1.1 What Are Peptides?

Peptides are organic molecules that consist of amino acids linked together in a polypeptide chain by peptide bonds. While they are often defined by the fact that they are the "building blocks" of proteins, their organic importance goes some distance beyond structural contributions. Peptides can serve as biological signals that mediate cell interactions and affect numerous physiological processes. That pathogenic versatility is a signature feature that sets them apart from other biomolecules, which makes them critical elements in both primary biological processes and high-level biotechnological applications.

Peptides: A definition and some introductory chemistry

Peptides are short chains of amino acids that bond with each other through peptide bonds that shape through a condensation reaction. The reaction comprises the carboxyl group (-COOH) of 1 amino acid and the amino group (-NH_2) of one other, releasing a water molecule and forming a covalent bond. The identity of the amino acids that make up a peptide (the number one structure) is a cornerstone of the trait since relatively simple alterations can re-tailor the biological activity of this molecule.

An amino acid, classically, is an organic compound of an alpha carbon atom bonded to an amino organization, a carboxyl organization, a hydrogen atom, and a one-of-a-kind side chain (R group). An introduction of their houses performs a vital characteristic of a peptide, which is decided via the kind and series of these aspect dressings, as this might also be polar, nonpolar, acidic, or basic. It allows peptides to take on various forms and interact with unique molecular targets, which helps to make them versatile in application.

Peptides are typically categorized according to length, with peptidomes containing less than 50 amino acid residues. This lesser size than that of proteins allows the peptides to perform motile functions as they appear as hormones, neurotransmitters, and antimicrobial dealers. In addition, their simple structure renders chemical synthesis feasible, thus opening up new possibilities for developing peptide-based therapeutics.

These are the fundamental differences between peptides and proteins.

Peptides and proteins have the exact composition of amino acids. However, they are typically discerned by their length and structural complexity. Proteins are giant macromolecules of 1 or more lengthy polypeptide chains that immediately filament into complex 3D structures. Stabilization of these systems occurs due to several interactions, such as hydrogen bonds, ionic interactions, van der Waals forces, and disulfide bridges. Protein folding is essential since its activity usually reverses to preserve a unique shape.

Peptides, conversely, usually lack the massive secondary, tertiary, and quaternary structures found in proteins and are much shorter. Proteins can play multiple roles, acting as enzymes, structural elements, or transport molecules, whereas peptides tend to be more specialized. Peptide hormones like insulin regulate glucose metabolism simultaneously, while antimicrobial peptides are protective in opposition to pathogenic microorganisms.

A significant difference is in their biosynthesis and degradation—translation and folding of proteins through ribosomes and post-translational modifications. Peptides, however, can be made through both ribosomal and non-ribosomal pathways. Non-ribosomal peptides often depict novel functionality due to their use of rare amino acids and cyclic architecture, which are synthesized by dedicated non-ribosomal peptide synthetase (NRPS) enzyme families.

Furthermore, peptides have shorter half-lives than proteins as they are more susceptible to enzymatic degradation. Peptides can operate this quickly because they have a speedy turnover themselves. That means that they provide a temporary signal, allowing the body to respond rapidly and flexibly to the variation in its physiology.

Not only structural but also functional differences exist between peptides & proteins. Though proteins commonly act as everlasting practical, peptides often function as short-term regulators or mediators, providing rapid communication within and among cells.

1.2 Different Types and Classifications of Peptides

Classification of peptides is mainly dependent on size, shape, and characteristics, as they also play diverse roles in biological structures. In this part, you discover a selected variety of these peptides, whereas inside, it is a framework for understanding their various uses and strategies of motion.

Oligopeptidos, polipeptidos and proteinas

The class of peptide can be grounded in the various types of amino acid residues they encode for. This category indicates the range of small peptides to large protein molecules.

Oligopeptides: are the most minor peptides and usually contain 1 to 10 amino acids. They frequently act as intermediates in metabolic pathways or signaling molecules. Dipeptides such as carnosine and tripeptides like glutathione, which are essential in antioxidant defense and detoxification of the cells, are examples.

Polypeptides: Polypeptides are extended chains of amino natural acids, nominally containing between 10 and 50 residues. They may either operate in isolation or as precursors to fully folded proteins. Polypeptides — for instance, adrenocorticotropic hormone (ACTH)- stimulate cortisol production in response to stress.

Proteins: Polypeptides with more than 50 amino acids and generally have the stable 3-dimensional configuration of the polypeptide are usually said to be proteins. Proteins perform a wide variety of functions, such as catalyzing biochemical reactions (enzymes), providing structural support (collagen), and mediating immune responses (antibodies).

This classification stresses the structural and functional variability in this peptide and protein family. Although the difference between peptides and proteins is strictly one of size, it also reflects differences in complexity, stability, and biological function.

Functional Classifications (Hormones, Antibiotics, etc.)

Peptides (stylized as Peptide) are more than just size; they are class based on functional characteristics in biological systems. This practical type illustrates their flexibility and critical function in retaining homeostasis and responding to environmentally demanding situations:

Hormonal Peptides are chemical messengers like insulin, glucagon, and oxytocin. They are produced by endocrine glands and secreted into the bloodstream, where they travel to target organs and modify diverse physiological functions. One such, for instance, is insulin, which helps control blood sugar levels by stimulating glucose uptake in the cells. In contrast, oxytocin can influence social bonding as well as reproductive ability.

Antimicrobial Peptides (AMPs): AMPs are critical to the innate immune system. They display broad-spectrum activity against bacteria, viruses, and fungi, typically disrupting microbial membranes. A few examples are defensins and cathelicidins, which protect epithelial surfaces against infection.

Neuropeptides: Peptides of substance P, endorphins, and vasopressin, which are unique among the peptides in the nervous system as neurotransmitters or neuromodulators. They modify multiple processes, such as pain perception, mood, and fluid balance. Neuropeptides regularly paint with classical neurotransmitters such as dopamine and serotonin, modulating their release and activity.

Peptide Enzyme Inhibitors Inhibition of enzyme hobby by some peptides makes them regulators of metabolic pathways or defensive retailers. An example is angiotensin-converting enzyme (ACE) inhibitors, a class of peptides utilized to treat hypertension by inhibiting the conversion of angiotensin I to angiotensin II, an extremely potent vasoconstrictor.

Such purposeful diversity indicates that peptides are adaptable to different physiological contexts. They are not limited to specific tissues or organs but broaden to almost all aspects of biological function.

1.3 Biological Functions of Peptides

Endowed with many biological functions, peptides usually mediate critical physiological processes. The role of hormones, neurotransmission, antimicrobial protection, and immune system modulation mediated by these molecules is a clear example of their fundamental functions in retention of fitness and homeostasis.

Hormonal Regulation

Peptide hormones are essential coordination factors that couple the organ systems of a sport. Synthesized in specific glands, they get released into the bloodstream in response to particular stimuli. After secretion, these hormones bind to receptors of course cells and initiate intracellular signaling pathways that modify cell excursion.

Insulin is an example of a peptide hormone secreted by the pancreas and regulates glucose metabolism by promoting glucose uptake into cells. This process is essential for keeping blood sugar levels in a very tight range. Likewise, the parathyroid hormone (PTH) controls calcium levels in the blood by increasing bone resorption and enhancing intestinal absorption of calcium.

The specificity of hormonal peptide signaling results in closely regulated physiological strategies. Dysregulation of peptide hormones can result in a wide range of complications, including diabetes, thyroid disorders, and growth defects, emphasizing their essential function in health and disease.

Neurotransmission

Neurotransmitters, neuromodulators, and positive peptides act in the worried device by influencing neural communication and behavior. Significant neuropeptides involved in the ache transmission and modulation methods are substance P and enkephalins. Substance P transmits pain signals from the periphery to the principal worried machine, while enkephalins serve as endogenous analgesics to suppress ache sensation (Naguib et al., 2001).

Furthermore, neuropeptides modulate emotional states and stress reactions. This is reflected, for example, in oxytocin—known as the love hormone and involved in social behavior and trust effects—as well as the strain response through the release of CRH, which triggers the release of ACTH from the pituitary gland.

This ability of neuropeptides to modify neurotransmitter secretion and neuronal activity highlights their importance in maintaining cognitive and emotional fitness. Neuropeptide signaling dysfunctions have been associated with a range of neurological and psychiatric disorders, including depression, anxiety, and schizophrenia.

Antimicrobial Properties

Abstract Antimicrobial peptides (AMPs) are one of the most essential components of the body and defense system, providing a rapid and effective response to microbial threats. These peptides exhibit wet antimicrobial activity duty on various pathogens, bacteria, viruses, and fungi.

The mechanism of action of AMPs is mainly based on the disruption of pathogen cell membranes, causing lysis and cell death. This method of movement is very different from regular antibiotics; therefore, AMPs are precious in the fight in the direction of antibiotic-resistant pathogens. Moreover, certain AMPs can modulate the immune reaction of their respective hosts to accentuate the recruitment and activation of other immune cells.

Indeed, the range of therapeutic use of AMPs spread beyond that of antimicrobial infections. Considering their broad utility in medicine, they are also being investigated for their role in most cancer remedies, wound healing, and treating inflammatory illnesses.

Immune System Modulation

They are also crucial in regulating the action of the immune response. Small tissues and proteins called cytokines and chemokines serve as cellular messengers, altering immune cells' activation, proliferation, and mobility. These peptides ensure the immune response is appropriately focused and sized to prevent excessive tissue injury or recurrent inflammation.

Cytokines such as interleukins and interferons are involved in the mediation of immune responses to infections and tumors. In contrast, chemokines direct immune cells to sites of illness or injury, helping eradicate pathogens and heal injured tissues.

The ability of the peptides to modulate immune function makes them especially attractive candidates for therapeutic development. Innovative scientific remedies for autoimmune ailments, most cancers, and infectious sicknesses are below way or being developed using peptide-based immunotherapies.

1.4 Peptide Structures

The generic shape of the peptide is paramount to its quality. The exceptional spans of peptide structure can be understood in more significant aspects of peptide stability, interactions, and proper biological sporting events.

Primary Structure

The primary structure of a peptide is its sequence of amino acids, dictated by the genetic code. The order of these dictates the chemical houses that make up the peptide and its biological activity. Deviations in the significant structure, even through a single amino acid, can tremendously affect the interest and specificity of the peptide.

For example, the differences between the active and inactive forms of positive peptides might be due to subtle alterations in their primary structure. The primary structure also determines a peptide's interaction with its target, which modulates its binding and therapeutic potency.

Secondary and Tertiary Structure

Peptides usually fold into unique 3-D conformations to perform their biological functions efficiently. The secondary structure encompasses well-known motifs of alpha-helices and beta-sheets, which may be stabilized by hydrogen bonding between the peptide backbone. These systems are part of the global stability of the peptide or peptidomimetic and can interfere with its binding to other molecules.

In some instances, the peptides continue to fold into a complex tertiary structure, which is held together by most bonding, such as hydrophobic interactions and ionic bonding (Some bonding is strengthened). These higher-order systems are necessary for peptides that would like to bind to their targets, receptors, and enzymes specifically.

The insight gained from such peptide systems not only helps to capture their biological functions but also helps to develop synthetic peptides for therapeutic applications. Changing the shape of a peptide allows researchers to adorn the balance, specificity, and interest of its garb of drug tendencies.

Chapter 2: Fundamentals of peptide synthesis

2.1 Introduction to Peptide Synthesis

The ability to create peptides in the lab has found its place as a stepping stone of modern-day chemistry and biochemistry, leading to advancements ranging from improved medicine to novel materials and technology. Peptide creation is the emergence of peptides at the core level. Still, maybe not everybody is mindful of these famous vehicle chains of creation in individual cells- little Western chains of amino acids linked with peptide bonds. These biomolecules are crucial in many biological processes, including enzymatic functions, cell signaling, and immune responses. Next to their importance, synthetic peptides have now become an essential element of research and therapeutic packages. This section narrates the early pathways of peptide synthesis, its growth over time, and the contemporary characteristics and techniques delineating this exciting field.

Brief History and Development

The history of peptide synthesis goes back to the late nineteenth and early twentieth centuries when Emil Fischer first established the structure of proteins and peptides. Early peptide bonds at Fischer's paved the way for synthetic peptide chemistry. Natural chemistry complemented basic studies on protein structure and biological activity, as by the mid-20th century, new methodologies had permitted the laboratory synthesis of simple peptides. The development efforts became exhausting in-depth and restrained in scope.

This query was answered by the development of solid-phase peptide synthesis (SPPS) by Robert Bruce Merrifield in the 1960s. This innovative method streamlined peptide assembly, enabling complex sequences to be synthesized automatically. For this work, Merrifield earned the Nobel Prize in Chemistry in 1984, paving the way for explosive growth in peptide science. In the following decades, device developments like protecting groups, coupling reagents, and automated synthesizers transformed peptide synthesis into a relatively fast and programmable apparatus.

Peptide synthesis plays a crucial role in the wide variety of packages, from peptide-primarily based drug improvement to diagnostic tools and biomaterials. Absorption These have garnered focus for significant years, with researchers bettering synthesis approaches around compelling originals, modeling tasks, and scalped real-time behavior generation.

Current Trends and Techniques

Peptide synthesis strategies have made apparent advancements in the last few years. One great fashion is the rising adoption of automatic synthesizers that automate portions of the synthesis manner, minimizing the time required to synthesize a molecule and the hazard of human mistakes. These systems have high-end software that optimizes response situations and tracks progress in real-time.

Microwave-assisted synthesis, which accelerates reaction rates and increases peptide yield, is another significant enhancement that has been embedded. This approach is beneficial for synthesizing long or complex sequences from steric difficulties and side reactions.

In addition, the researchers also developed the green chemistry approach to peptide synthesis by reducing waste and using fewer hazardous reagents. The innovations, including solvent-free coupling reactions and recyclability of this catalyst, contribute to the global sustainability trend in chemical research.

Such peptide synthesis likewise expands into non-natural peptides of altered amino acids, cyclic peptides, and peptide-mimetic systems. These molecules have been shown to possess improved stability, bioavailability, and specificity over other macromolecules and are helpful in drug discovery and biomaterials development.

2.2. Solid-Phase Peptide Synthesis (SPPS)

SPPS is the most commonly used method for peptide synthesis; this fact is of great relevance to our study. Aided by Robert Bruce Merrifield, SPPS transformed the field by allowing peptides to be synthesized stepwise and anchored to a solid support. That comes with many advantages over traditional liquid-section synthesis: easier purification, automation potential, and suitability for complex sequences.

Overview of the SPPS Process

In the SPPS manner, the first amino acid is linked to a strong resin, supporting further reaction steps. To prevent unwanted side reactions, the amino institution of the connected amino acid is blanketed. SPPS consists of a repeating cycle of the following steps: the deprotection of the amino functionality, the coupling of the following amino acid, and the washing away of unreacted reagents.

After identifying the preferred series of peptides, the peptide must be cleaved away from the resin and purified. This step often requires the deprotection of side-chain protecting groups to produce the final product. SPPS is also especially suitable for synthesizing long and complex peptides because of its iterative nature.

Materials and Reagents

Abstract SPPS stands for solid-phase peptide synthesis based on several specific materials and reagents. That stable guide, or resin, is the anchor point for the growing peptide chain. The most common resins are polystyrene and polyacrylamide, which are functionalized to enable covalent coupling of the first amino acid.

Materials and Methods Protecting agencies were employed to arrest reactive practical corporations on amino acids, ensuring that separation coupling reactions proceed selectively. Deprotection in organic synthesis is the reverse of protection and involves the removal of one or more protecting groups from the molecule or (polymer); common protecting agents include fluorenylmethyloxycarbonyl (Fmoc) and tert-butyloxycarbonyl (Boc). This is especially true for Fmoc-primarily based SPPS because the deprotection conditions are moderate.

Coupling reagents enable the formation of peptide bonds between amino acids. Representative examples include the carbodiimides (e.g., DCC, DIC) and uronium salts (e.g., HBTU, HATU). These reagents induce the arrangement of a carboxyl of the assaulted amino acid that favors nucleophilic attack by the amine group of the growing peptide chain.

SPPS Pros and Cons

SPPS provides many blessings over its preceding strategies. Utilizing a solid aid purification is simplified, where unreacted reagents and with-the-aid-of-products can be eliminated by washing the resin. Combined with the fully automated approach facilitated by SPPS, this attribute allows extremely pure peptide synthesis.

However, SPPS has its challenges of its own. In aspect reactions, for example, racemization or incomplete reactions might also be shown due to the repetitive deprotection and coupling cycles. These problems may diminish the production and quality of the final peptide. The solubility and stability of the generated peptide library also limit the process and induce constraints on this method.

2.3 LPPS (Liquid-Phase Peptide Synthesis)

Although SPPS is predominant in the area, liquid-phase peptide synthesis (LPPS) remains a valuable method for selected applications. In LPPS, peptides are assembled in solution, offering several advantages and challenges compared to SPPS.

Intro + a Simple Process

Peptide chains target the successful assembly of an answer-segment chain in LPPS. Each step involves coupling one amino acid to the N-terminus of the growing peptide chain, monitored using techniques such as purification to remove side products and unreacted reagents. While the former utilizes robust facilitation, the latter sits on soluble intermediates for separation and purification instead of SPPS.

LPPS typically involves small peptide synthesizing or vast parts of the targeting arrangements. It is also desirable to synthesize peptidomimetics that are difficult to synthesize on a solid support, such as those containing hydrophobic and aggregation-prone sequences.

Key Differences from SPPS

One of the main differences between LPPS and SPPS is the presence of a strong theory. The peptide is removed from the soluble phase at the end of the synthesis process. However, we need extensive purification after each step in LPPS. LPPS is more extraordinary in-depth exertion and time-ingesting than SPPS due to this characteristic.

LPPS offers more versatility in response scenarios and solvent needs, though. This flexibility can be high-quality for challenging sequences or for accelerating unique reactions. Further, LPPS allows easier access to longer peptides than SPPS permits, typically due to the steric hindrance associated with SPPS.

2.4 Peptide Bond Formation

The peptide bond formation is central to peptide synthesis and is the fundamental chemical transformation that connects amino acids into chains to form peptides. This cytoskeletal system is essential for organic and synthetic environments and embodies the structures and roles of proteins and peptides. Complete cognizance of the chemistry and mechanisms of peptide bond formation is paramount to enhancing artificial methods to yield excessive yield and purify and constant peptides.

Release of By-products and formation of Peptide

The last step involves the excretion of the waste product, which is urea, alcohol, or some other compound, depending on the coupling reagent used. Therefore, The peptide bond proceeds in the preferred orientation, and the reaction progresses through its subsequent cycles to further elongate the peptide chain.

Role of Coupling Reagents

Peptide bond formation is energetically unfavorable (Peptide bond formation is energetically unfavorable.); therefore, coupling reagents are required during peptide synthesis to drive this chemical reaction. Selecting the appropriate coupling reagent is significant as it instantly influences the response efficiency, cost, and selectivity. Multiple training of coupling reagents were decided to pull aside, every with experience and uses.

2.5 Enhancement of Peptide Bond Synthesis

The reaction conditions must be carefully optimized to achieve high yields and purity of the synthesized peptide. The choice of solvent, temperature, response time, and the interest of reagents are essential components that influence the success of the coupling reaction.

Solvent Selection:

During peptide synthesis, the solvent must dissolve all reactants and reagents and be palatable to keep the intermediates stable. These industrial solvents include dimethylformamide (DMF), dichloromethane (DCM), and N-methylpyrrolidone (NMP). The solvent used may affect the reactivity of the coupling reagent and the solubility of the growing peptide chain.

Nervous System and Reaction Time & Temperature:

High temperatures tend to improve the reaction cost; however, they may enhance the danger of side reactions, including racemization. On the other hand, higher temperatures reduce the risk of facet reactions; however, they can sluggish the response at the same time. Green Peptide Bond Formation Related to the Right Stability Typically, the instances for response range from a few minutes to some hours, depending on the process and technique used.

Racemization and Coupling Efficiency:

The obverse of this is the racemization of the chiral center of an amino acid, which represents one of the significant challenges in peptide synthesis. Racemization culminates in the preference of incorporating D-amino over L-amino acids, forming peptides with altered organic function and decreased purity. Optimized coupling reagents, mild reaction conditions, and reagents such as Oxyma can significantly reduce the likelihood of racemization.

Elimination of Side Reactions and By-products:

The occurrence of side reactions, as well as the formation of diketopiperazines or aspartimides, can occur throughout peptide synthesis, especially when synthesizing sequences containing proline or aspartic acid. V Yields of target products are decreased when these facet reactions occur, complicating purification. Careful control of reaction conditions and special additives could alleviate those problems.

Technological Improvements in Peptide Bond Formation

Recent developments in peptide synthesis have focused on the greenness and efficiency of peptide bond formation. One of these is microwave-assisted peptide synthesis, a technology that has received significant uptake. Microwave irradiation enhances response prices, lowering workflow time and enhancing expected yield. This approach is beneficial for synthesizing long or challenging peptide sequences.

We also have promising methods, such as drift chemistry, where reactions are performed in a continuous flow device instead of batch mode. Flow chemistry mainly influences response parameters, resulting in improved reproducibility and scalability. It also minimizes waste and allows the use of reagents in an environmentally friendly manner, following green chemistry concepts.

One of the new fields of peptide synthesis is enzymatic, which utilizes the catalytic power of enzymes to create peptide bonds. Proteases and ligases catalyze the formation of peptide bonds with high specificity and few side reactions at low temperatures. This possibility also seems perfect for producing complicated peptides and proteins with post-translational modifications.

Role of the Formation of Peptide Bonds in Peptide Functions

One of the significant factors that simultaneously influence the activity and functionality of synthetic peptides is the quality of the peptide bond formation technique. Peptides have a wide range of pharmaceutical, biotechnological, and industrial applications. Peptide-based total pills such as insulin, glucagon, and somatostatin rely on precise peptide bond formation for their healing effectiveness.

Artificial peptides work as probes, antigens, and enzyme substrates in biotechnology. These are essential for those applications because slight differences can influence binding affinity, pastime, and specificity, as well as the accuracy of the peptide bond formation. In a similar context to substance technology, peptides play a role in developing biomimetic substances and nanostructures. The integrity of the peptide bonds directly influences their mechanical and purposeful property.

Peptide bond formation is an indispensable step in peptide coupling, and attention must be paid to response processes, coupling reagents, and synthesis conditions. Improvements in coupling technology, fulfillment in optimization of coupling and chain elongation procedures, and state-of-the-art methods in synthesizing oligomers preserve the efficiency and scope of peptide synthesis. With knowledge of peptide bond formation, scientists and chemists can generate excellent peptides for various applications — driving advances in medicine, biotechnology, and more.

Chapter 3: Peptide Clean-Up and Characterization

Peptide purification and analysis is an indispensable system in many disciplines, and alertness from basic biochemical studies to high purposes is restorative development. Chemical or biological synthesis of peptides usually provides a complex mixture containing the desired peptide and impurities such as truncated sequences, deletion sequences, or side products. Those peptides' isolation and subsequent testing are paramount to guarantee the final product's functionality, safety, and effectiveness.

This chapter describes the methods and strategies employed to purify and analyze peptides. Furthermore, this builds upon 3 critical topics in the dialogue: methods for peptide isolation, analytical techniques in validating structure and structure, and hints that arise between synthesis and isolation. Every phase provides a complete description of those techniques' theoretical concepts, instrumentation, and practical use.

3.1 Methods for Peptide Purification

Peptide purification is essential in synthesizing peptides, aiming to obtain the final product as pure as possible, free from impurities such as truncated sequences, side products, or other contaminants that may arise during synthesis. Purification is not only to obtain the target peptide but also to get the maximum yield and purity. They use a chromatographic technique, each specific for the individual properties of the peptide concerned and the impurities gift. High-Performance Liquid Chromatography (HPLC), Gel Filtration, and Ion-Exchange Chromatography are the most commonly used strategies for purifying Peptides. While these strategies vary in their ideas of separation, programs, and being concerned with performance, they provide a comprehensive toolbox for researchers and manufacturers wishing to reap the desired purification.

Polishers at LHR (Liquid High-Performance Liquid Chromatography).

High-Performance Liquid Chromatography (HPLC) HPLC has arguably been the most versatile and widely used method for purifying peptides. Due to its ability to discriminate peptides mainly based on slight differences in their chemical addresses, including hydropathy and polarity, is an essential tool in research and industry. This HPLC requires forcing a liquid sample through a column packed with a stationary phase, which interacts differently with different sample components and causes their separation.

Reverse-Phase HPLC (RP-HPLC)

The most common mode of HPLC for peptide purification is RP-HPLC. In this method, the stationary phase is nonpolar, usually consisting of silica particles coated with hydrophobic chains such as C18 or C8. However, the cellular segment is polar: water and an organic solvent such as acetonitrile or methanol. During the passage of the peptide sample through the column, peptides that contain more hydrophobic residues interact more extensively with the stationary phase and elute later. In contrast, far less hydrophobic peptides will elute more quickly.

In RP-HPLC, gradient elution is generally employed, in which the quantity of the organic solvent in the mobile phase is gradually increased. This approach improves separation by reducing interactions between the peptides and the desk-bound phase in a stepped manner, and it has the capability of peptide separation with very similar hydrophobic characteristics. The gradient profiles, float prices, and column temperatures can be fine-tuned to improve the separation process significantly.

Normal-Phase HPLC (NP-HPLC)

Special programs use NP-HPLC; however, RP-HPLC is widely used for peptide purification. NP-HPLC uses a polar stationary phase (e.g., silica gel) and a nonpolar mobile phase. This configuration benefits the explicit separation of polar functional groups such as those seen in peptides or peptide derivatives. Due to the relatively limited set of conditions in which peptides can be extracted using nonpolar solvents, it is less frequently used as a strategy.

Detection Methods in HPLC

HPLC is frequently used with susceptible detectors that allow real-time detection of peptide purification. Thus, peptide bonds absorb at 214 nm or aromatic residues at 280 nm, widely used as UV detectors. Fluorescence detectors provide increased sensitivity, limited to peptides labeled with fluorescent tags. Since mass spectrometry (MS) detectors give the benefit of imparting molecular weight data in tandem with chromatographic separation, it is possible to identify peptides with high specificity throughout purification.

While HPLC has many advantages, it also has some challenges. However, large-scale packages could contain challenges due to costly and environmentally hazardous organic solvents. Moreover, the price of the HPLC system and its preservation ability is one of the causes for small laboratories.

Gel Filtration

An equally important strategy for peptide purification is gel filtration, or size-exclusion chromatography (SEC). This technique divides peptides primarily based solely on the molecular length and is, therefore, different from HPLC, which relies on chemical residences such as polarity or rate. The column is packed with porous beads made of dextran, agarose, or polyacrylamide in gel filtration. These beads have specific pore sizes, allowing small molecules to enter while keeping larger molecules out.

Principles of Separation

The separation principle is simple in gel filtration. Smaller peptides take longer routes through the column as they penetrate the beads' pores, thus eluting later. Larger peptides are excluded by a cut-off, meaning they cannot enter the pores and are eluted first. Since gel filtration is size-based total separation, it's miles specifically appropriate for eliminating small-molecule impurities (e.g., salts, solvents, or unreacted reagents) from peptide solutions.

Applications and Advantages

Typically, gel filtration is used as a secondary purification step after HPLC for desalting, buffer change, or small contaminant removal. This low-destructive provisional method reinforces touchy peptides without dismaying their structure. This renders it more appropriate for heat-sensitive peptides that aggregate, denature, or degrade.

Gel filtration is also simple and does not require high technical skills. It avoids time-consuming sample preparation before extraction and is easy to elute. Gel filtration columns can be reused multiple times with appropriate cleaning and storage, making the method economical for repeated applications.

Limitations

Its shallow resolution compared to HPLC is the major limitation of gel filtration. This implies that peptides with similar MW can not be adequately fractionated. For low-concentration peptide samples or complex mixtures with an extensive dynamic range in sizes, there are many problems with using gel filtration due to loss of resolving power. Additionally, the method is slower than HPLC as it relies exclusively on gravity or low-pressure pumps to move the sample through the column.

Ion-Exchange Chromatography

Ion-exchange chromatography categorizes peptides according to their fee properties, providing a complementary approach to HPLC and gel filtration. In this method, a stationary segment with politically evoked regions interacts with reverse-charged groups of peptides. The fierce or awful habit of the peptides on the functioning pH will determine the kind of ion-change resin used.

Cation-Exchange Chromatography

In cation-change chromatography, the desk-bound segment incorporates negatively charged agencies that bind undoubtedly charged peptides (cations). Peptides with a higher internet superb price bind to the resin more strongly, while peptides with a lower or neutral fee elute earlier. There is a driven elution of sure peptides by elevating the ionic power of the mobile section or by adjusting pH to decrease the internet-positive charge of the peptides.

Anion-Exchange Chromatography

In contrast, anion-change chromatography occurs on a static phase carrying positively charged entities that interact with negatively charged peptides (anions). It separates and elutes peptides based on their net negative charge, similar to a cation swap. This has been a boon for acidic peptides that bear a terrible price at physiological or primary pH tiers.

Optimization Strategies

Ion-change chromatography is a technique whose success depends on fine-tuning experimental conditions. Factors including buffer selection, pH, ionic power, and gradient profile are critical for efficient separation. For example, the pH of the buffer must be carefully selected to ensure that the target peptide and impurities have different charge states. Similarly, the cell segment's ionic energy should be optimized to fine-tune the peptides' binding and elution.

Applications and Benefits

Both analytical and preparative purposes are extensively purposed by ion-trade chromatography. This makes price houses particularly powerful for purifying peptides with closely related sequences. The latter comprises isolating isoforms, submit-translationally modified peptides, or peptides with minor collection versions. The strategy can also be scaled up and down, so it can be applied to small-scale research or large-scale manufacturing and all steps in between.

Challenges and Drawbacks

Ion-change chromatography has some limitations despite its advantages. Minor adjustments in buffer conditions easily poison the approach, whereas misguided optimization will result in dreadful decisions or peptide loss. Also, binding interactions in ion-alternate chromatography are pretty robust, allowing for elution of peptides needing high salt concentrations or pH values. Not all peptides will be compatible with these conditions, especially aggregation-prone or denaturing peptides.

A Comparison Among Different Purification Methods

The choice of purification method depends on the peptide's unique properties and the final application. Although a peptide can easily be purified by various means, high-resolution HPLC is often the first resort for purification, particularly for complex combinations or where a high degree of purity is required. Gel filtration is an excellent complementary method to remove salts and low molecular weight contaminants, whereas ion-exchange chromatography provides a robust alternative to peptide separation based on charge.

Each of these approaches has its strengths and its barriers. Although HPLC provides an unprecedented resolution, it comes with a higher cost and with organic solvents. Although gel filtration is gentle and easy to use, its resolution is insufficient for closely related peptides. Ion-alternate chromatography is particularly useful for purifying peptides with minor differences in their rate but requires careful optimization of buffer conditions.

The merger of those strategies can allow researchers to return excessive tiers of purity and yield so that the purified peptides fulfill the excessive criteria needed for examination or within the diagnostics and healing functions. In keeping with this, subsequent improvement of these purification techniques is thus an indispensable talent for any scientist working within the region of peptide chemistry.

3.2 Analytical Techniques

It's the easiest step in purifying peptides that guarantees their utility and effectiveness in scientific research or their therapeutic packages. Peptides must then be characterized to identify them, determine their purity, and verify purity characteristics. Research methods provide information regarding the series, conformation, and standard quality of the peptide(s) to guarantee that the produced peptides satisfy the standard specifications easily. Mass Spectrometry (MS), Nuclear Magnetic Resonance (NMR), and Amino Acid Analysis are some of the most potent and applied methods to assess peptides. All three methods provide information on the peptide in question, which leads to comprehensive knowledge of the properties of molecules.

Resolving Sequences by Mass Spectrometry (MS)

Since its first derivation, mass spectrometry (MS) has opened a new dimension in peptide assessment. It is presently becoming part of any peptide characterization process due to its unparalleled sensitivity, specificity, and accuracy and being an extensively specific molecular platform. Determination of Molecular MassDetermination of Molecular MassMS is undoubtedly one of the most effective strategies for analyzing the molecular weight of peptides. It also serves as the primary performer of MS in peptide evaluation. Furthermore, it provides information on the sequence of amino acids, identifies submit-translational modifications, and allows the detection of impurities.

Ionization Techniques in MS

Mass spectrometry(MS), commonly used in peptide analysis, is based on two ionization techniques called electrospray ionization(ESI) and matrix-assisted laser desorption/ionization(MALDI).

Electrospray Ionization (ESI) — This method forms ions by applying high voltage to the liquid that contains the peptide sample as it passes through a delicate pull. Its effects create charged droplets that evaporate, giving the peptide ions. ESI is particularly suited for analyzing larger peptides and proteins because it generates multiply charged species, allowing the analysis of higher molecular weight peptides within the mass range of the instrument.

1) Matrix-Assisted Laser Desorption/Ionization (MALDI): MALDI uses the energy of a pulse laser to extract peptides from a matrix through ionization. When the laser hits, the matrix absorbs some of the laser power to aid peptide desorption and ionization. This method is simple, robust, and valuable for studying complex peptide combinations and is the most commonly used in the field. Unlike ESI, MALDI tends to generate singly charged ions, making spectral interpretation easier.

MS/MS

One central virtue of MS in peptide examination is its capability to perform tandem mass spectrometry (MS/MS). In MS/MS, the peptide ions of interest are fragmented in the mass spectrometer to provide smaller ion fragments. While analyzing these fragments, one can retrieve information regarding the peptide pool, as each fragmentation event generally corresponds to the breakage of a particular bond within the peptide backbone.

MS/MS is instrumental in:

Determination of sequence: The sequential loss of amino acids during fragmentation enables one to reconstruct the unique sequence of a peptide.

Post-translational modification (PTM) analysis: MS/MS can identify modifications such as phosphorylation, glycosylation, or acetylation based on mass shifts (covariance of charge 20 or 40 Da).

They identify and characterize impurities or by-products of synthesis via their exact mass spectra or Impurity profiling.

Applications and Limitations

Mass spectrometry (MS) has a high sensitivity that enables the detection and analysis of nanogram peptides to picogram quantities, which is beneficial in proteomics and pharmaceutical studies. In addition, MS can be coupled with chromatographic techniques such as High-Performance Liquid Chromatography (HPLC) to facilitate the rapid separation and analysis of more complex peptide populations.

Despite this, MS does have its drawbacks. It also may require specific systems and data sets, and converting spectra, especially for complex peptides or mixtures, can prove challenging. Further, peptide composition can influence ionization efficiency, thus resulting in biased results.

Nuclear Magnetic Resonance (NMR) for Structural Analysis

MS generates molecular weight and sequence information par excellence, but Nuclear Magnetic Resonance (NMR) spectroscopy provides unparalleled information about the 3-dimensional conformation and dynamics of peptides. NMR is a non-destructive two-detailed analytical method that exploits the magnetic properties of native atomic nuclei to obtain mom data about structure.

Principles of NMR

NMR takes advantage of the fact that some atoms, like hydrogen (^1H) and carbon (^13C), have intrinsic magnetic moments. If those nuclei are placed in the strong magnetic field, they will orientate with and against the field, resulting in two distinct energy levels. The nuclei can be excited by higher power states by radio wave pulses. The nuclei emit the same energy when returning to their lower energy states. This energy is emitted as the nuclei return to lower energy states, which is detected and converted to an NMR spectrum.

Uses of Peptide Testing

NMR is handy for reading peptides in solution and provides information that includes:

Secondary and tertiary: An NMR can pick up on α-helices, β-sheets, and random coil regions in peptides.

Conformational dynamics: The ability of peptide chains to be flexible and move can be evaluated, providing helpful knowledge on their functional houses.

NMR enables studies of peptide interactions: Peptides typically display conformational flexibilities within their structure and in their interactions with proteins, nucleic acids, or small molecules—an essential part of drug design and understanding organic pathways.

Two-Dimensional NMR (2D NMR)

One-dimensional NMR spectra are crowded and difficult to interpret for complex peptides. Two-dimensional spectrometries, like COSY (Correlation Spectroscopy), NOESY (Nuclear Overhauser Effect Spectroscopy), and HSQC (Heteronuclear Single Quantum Coherence) provide better separation and allow researchers to know about positional correlations in atoms.

For instance:

- COSY and TOCSY provide information on scalar coupling, which helps distinguish the peptide's spin systems.
- As a start, NOESY shows how close some atoms are in space, which is needed to solve its 3D shape.
- HSQC - Interprets the hydrogen and the carbon nuclei, Which provides the chemical imprints of the peptide

Advantages and Limitations

This unique capability to supply structural data in an atomic-degree resolution in an answer makes NMR indispensable to structural biology. Unlike X-ray crystallography, NMR does not require that peptides twist themselves into crystals and, therefore, can be observed in a more physiologically relevant environment.

Yet, NMR has benefits and challenges. It is much less sensitive than MS, requiring more significant numbers of patterns, typically within the millimolar range. Moreover, there is an inverse correlation between the size of the peptide and the dimensions of the NMR spectra, which becomes increasingly complex for large peptides or proteins, posing a substantial challenge to data analysis.

Amino Acid Analysis

Although classical, the analysis of amino acids represents an efficient approach to confirming peptide composition and identity. This approach involves hydrolyzing the peptide into its amino acids and measuring them to confirm its sequence and composition.

Hydrolysis of Peptides

Peptide hydrolysis The general amino acid analysis procedure is always initiated with peptide hydrolysis. This process usually uses vigorous acidic or enzymatic conditions to cleave the peptide bonds and liberate individual amino acids. Although precise conditions may vary primarily based on the composition of the peptide and the amino acids of the hobby, 6N hydrochloric acid is the most prevalent technique of acid hydrolysis.

Quantification of Amino Acids

Posthydrolysis, the released amino acids are differentiated and measured using chromatographic strategies, including ion-exchange chromatography or opposite-phase HPLC. The amino acids are frequently derivatized to increase detection sensitivity, and standard derivatization techniques include phenyl isothiocyanate (PITC) or ortho-phthalaldehyde (OPA).

Peptide analysis applications

Amino acid analysis has more than a single utility:

- Peptide composition verification: Matching the measured amino acid composition to the theoretical values can demonstrate peptide identity and purity.

- Quality control guarantees that the peptide synthesis is successful and the final product satisfies the required specifications.
- Inference: This information furnishes the molar extinction coefficient, which is invaluable for electrolyte quantity of peptide concentrations in answer.

Advantages and Limitations

One of the reasons that amino acid evaluation is a hallmark of exceptional manipulation in peptide synthesis is that it's so exact and reliable. Even trace levels of positive amino acids can trigger it, giving it a complete peptide breakdown per se.

However, this method is time-consuming and requires careful handling to prevent the unwanted loss or denaturation of amino acids during hydrolysis. In combination with tryptophan and cysteine, some amino acids are sensitive to acidic environments and may degrade. Therefore, particular hydrolysis protocols are required to avoid the destruction of these residues.

3.3: Troubleshooting and Optimization

The yield and purity of peptide synthesis and purification processes regularly face multiple challenges. This is one of the emerging research areas that, when spotted and solved, can help work towards optimizing the process to produce great peptides. In this stage, you discuss routine synthetic problems and solutions and provide pointers for successful purification.

Synthesis Problems (and How to Solve Them)

Peptides synthesized through solid-segment peptide synthesis (SPPS) are susceptible to incomplete reaction, side reaction, and resin aggregation. Incomplete coupling or deprotection steps can result in truncated sequences, whereas side reactions such as oxidation or racemization can modulate the properties of the peptide. Most often, it is necessary to optimize the response situations, which might consist of using additional reagents, adjustments to coupling instances, or using more excellent energetic activating retailers to deal with the aspects related to these troubles. Methods such as HPLC or MS to monitor synthesis develop those issues early and allow gaps as much as possible.

Essential Tips To Get More Out Of Your Purification

This gimmick development calls for detailed full-size know-how of the physicochemical homes of the peptide, as well as the contaminants that need to be rejected throughout purification. This process includes the selection of the chromatographic method, gradient optimization, and optimal buffers required for purification. This separation level is improved with a combination of purification methods, along with RP-HPLC, by way of ion-exchange chromatography, enabling the fractionation of complex mixtures of peptides. Moreover, maintaining rigorous, satisfactory management measures, including each day calibration of gadgets and verification of approaches, ensures that purification outcomes are constant and constant.

Chapter 4: Peptides in Action: Real-World Applications

Peptides today have come a long way in technology, spanning many industries and fields. These versatile molecules, made up of short chains of amino acids, have proven advantageous in fields ranging from medicine and cosmetics to diagnostics and agriculture. Because artificial and natural peptides have become more significant and more reachable because of improvements within the biotechnology field, their sensible applications have accelerated exponentially. The chapter provides a more in-depth analysis of the central regions where peptides are transformative: therapeutic, beauty, diagnostic, and biotechnological.

4.1 Therapeutic Peptides

The roles of peptides are crucial additives in modern medicinal drugs because their particular houses make them extraordinary from conventional drug cures. These are helpful brief sequences of organic acid groups, which function as organic messengers, regulating various physiological processes. Due to their specificity, efficacy, and relatively low toxicity, they are relatively attractive therapeutic sellers. Unlike traditionally prescribed medication, which generally affects many receptor molecules and may additionally convey accidental aspect effects, peptides can be strategically created to interact with unique molecular targets. This accuracy allows for extra-targeted remedy methods, enhancing healing effects while reducing adverse reactions.

Peptides as Drug: The Mechanism of Action and Benefits

Peptides attach to specific cell surface receptors to induce or suppress a targeted biological response. This ability to emulate or intervene in natural organic strategies represents the basis of their applicability to a range of diseases. The focus on peptides provides one of the significant advantages, i.e., their high affinity and specificity for their targets. Such features best enhance therapeutic efficacy and reduce the likelihood of typical off-target results related to small-molecule capsules.

Peptides have an extraordinarily brief 1/2-life in the body, which is a hassle in some instances; however, they can be effective at other times. Because of this quick degradation setup, they're less likely to accumulate in tissues, thus reducing the risk of long-term toxicity. Extensive improvements in peptide engineering and ameliorating artificial analogs and modifications have also tackled some of the difficulties regarding their stability and bioavailability. These developments have set the stage for the successful use of peptides in diverse therapeutic areas.

Peptides in Cancer Therapy

Cancer continues to be one of the leading causes of death worldwide, and the identification of effective therapeutic strategies is an essential area of biomedical research. Peptides have proven to be a new class of therapy agents in oncology, presenting new alternatives for cancer diagnosis and treatment (1). They are located to play a multi-purpose position in cancer remedy applications as diagnostic markers, drug vendors, and direct healing providers.

Using peptides as carriers of cytotoxic agents is among the most exciting applications of peptides in cancer treatment. These conjugates (PDCs) are developed to deliver chemotherapeutic drugs directly to cancer cells, reducing the harmful effect on the whole body and increasing the therapeutic effect of the drugs. Cytotoxic compounds are conjugated to peptides that preferentially bind to receptors over-expressed on tumor cells, assuring the delivery of drugs at the tumor site while sparing normal tissues.

Peptide-primarily based vaccines are every other novel strategy in most cancer therapy. These vaccines train the immune system to recognize and target cancer cells. These generally include tumor-associated antigens that produce a robust immune response to inhibit tumor growth and prevent metastasis. Peptide vaccines have demonstrated this technique's ability to treat cancers such as melanoma, breast cancer, and prostate cancer. The effects of promoting safety and effectiveness have been validated in scientific trials.

Apart from acting as companies and vaccines, peptides also act as a minimum of direct inhibitors of cancer cell growth. Some peptides have been designed to interfere with signaling pathways necessary for tumor maintenance and development. Peptides that inhibit the activity of vascular endothelial growth factor (VEGF), for example, can suppress angiogenesis, the process by which tumors form new blood vessels to feed their growth. Those peptides cut the blood supply for the tumor, and most cancer cells can not spread without blood.

Peptide-based Medicines in the Management of Diabetes

Peptide-based therapeutic approaches have shown significant advancements in the treatment of diabetes, especially type 1 and type 2 diabetes. For almost a century, insulin (peptide hormone) has been used as a key to treat diabetes. But thanks to newer advances in peptide therapeutics, several options exist for developing treatments for this longtime ailment.

Insulin analogs, which are modified forms of insulin, provide advanced pharmacokinetic traits compared to regular insulin. Odumetion analogs aim to mimic the body more closely regarding the model of herbal insulin secretion, presenting superior glycemic management and a nominal risk of hypoglycemia. Insulin glargine and insulin degludec are long-acting shed insulins that release insulin over a long period and reduce the frequency of these typical injections as high as morbidly obese. In contrast, rapid-acting analogs such as insulin lispro and insulin aspart are rapidly absorbed, allowing for improved postprandial glucose control.

In addition to insulin, the development of glucagon-like peptide-1 (GLP-1) receptor agonists has transformed the management of type 2 diabetes. These peptides increase insulin secretion in response to elevated blood glucose concentrations, inhibit glucagon release, and delay gastric emptying to promote glycemic control. Today, GLP-1 agonists like exenatide and liraglutide have collectively also been demonstrated to induce weight loss, a key advantage for a large portion of type 2 diabetes patients who struggle with obesity.

Apart from their glucose-reducing effects, GLP-1 agonists have documented cardioprotective benefits, reducing the risk of major adverse cardiovascular events in high-risk patients. In conclusion, this dual benefit of glycemic and cardiovascular protection highlights the potential of peptide-based therapeutics in targeting the complex challenges of diabetes management.

Peptides in Antiviral Treatment

The global perspective of viral infection, which entails the human immunodeficiency virus (HIV), the influenza virus, and the various hepatitis viruses, warrants the ongoing development of therapeutic measures to combat these infections. What makes peptides so unique is they offer an entirely new method of treating viral diseases, targeting various stages of the viral life cycle to prevent infection and replication.

Peptides are most effective at inhibiting viral entry and are one of the number one pathways by which they enact their antiviral consequences. This is completed by enhancing mixture inhibitors, which obstruct the cooperation between viral envelope proteins and host cell receptors. An example is enfuvirtide, a peptide used to treat HIV, which works by blocking the fusion of the virus into the host cell membrane, preventing the infection process.

Peptides already have a role in blocking viruses' replication and assembly process. Instead, peptide-primarily based inhibitors target viral enzymes such as proteases and polymerases and can block the formation of recent viral particles. For example, this method has been examined throughout the event of medicines for hepatitis C, wherein peptide inhibitors propose the NS3/4A protease, an essential enzyme within the viral replication cycle.

Besides their direct antiviral effects, these peptides can adjust the immune responses to enhance the body's ability to clear viral infections. Some peptides function as immunomodulators, activating the production of antiviral cytokines and boosting the function of immune cells such as natural killer (NK) cells and cytotoxic T lymphocytes. Such an immunomodulatory effect is especially useful in chronic viral infections characterized by immune suppression or dysregulation.

Such versatility of peptides in antiviral therapy is also demonstrated by developing peptide-based vaccines against other viruses. These vaccines generate a targeted immune response against viral antigens, protecting against fatal infections. Peptide vaccines are now being researched for other viral threats, including COVID-19, thus vitalizing peptides in global fitness issues.

Overcoming challenges and future directions in peptide therapeutics

However, despite the considerable advances in peptide therapeutics, some challenges must be conquered to realize their full potential. A significant limitation of peptide tablets is their vulnerability to enzymatic degradation, which could reduce their bioavailability and therapeutic potency. The complexity requires us to develop systems for shipping peptides or to modify them to improve their stability and 1/2-life within the body.

This and the advent of drug shipping technologies like nanoparticles-primarily based totally structures and conjugation with polyethylene glycol (PEG) have validated promise in addressing the demanding situations. These strategies do not solely protect peptides from an enzymatic breakdown but also enhance their solubility and focus on specificity.

The peptide synthesis and manufacture cost is also an endeavor and can be pretty expensive when put next to small-molecule capsules. New green and cost-effective manufacturing techniques, including recombinant DNA technology and solid-phase peptide synthesis, must be developed to ensure peptide therapeutics are accessible.

Peptide therapeutics and their future destiny of peptide therapeutics remain ahead, with examples extending from preliminary explorations of new or investigational peptide sequences and characteristics. Artificial intelligence and computer-aided design can potentially speed up inventive new therapeutic peptides with improved properties. Moreover, combinations of peptides with various therapeutic modalities, including monoclonal antibodies and small molecules, could also exert synergistic effects and expand the therapeutic potential of peptide-based therapies.

With increased research in this field, peptides are set to play an even more significant role in tackling some of the most urgent health challenges we face today. Peptides have substantial healing power from most cancers and diabetes to viral infections and beyond, making it a new division of precision medicine and bettering patient results.

4.2 Cosmetic Peptides

The cosmetics industry has seen a super transformation with the emergence of bioactive peptides that offer a new targeted strategy to handle pores, skin health, and aging: peptides, short chains of amino acids, and clave as signaling agents in the body. The penetration capacity and the properties of modulating organic strategies promote their importance within the skin care field. In contrast to various lively elements, Peptides offer centered answers by stimulating precise molecular pathways. This trait has established them as a staple in developing advanced cosmetic treatments for skin rejuvenation, hydration, and anti-aging.

How Peptides Work in Your Skincare Routine

Peptides preserve the skin's fitness by influencing the manufacturing and reaction of essential proteins such as collagen, elastin, and keratin. Primarily, collagen helps our skin with elasticity and structural honing. Collagen loss is a normal process, the decline of which occurs as you age. Collagen Loss leads to visible signs & symptoms of getting old, Like — wrinkles, excellent traces & sagging pores and skin. The aging process is accelerated by collagen damage with exposure to external factors such as sunlight (UV), pollution, and lifestyle choices such as smoking and alcohol consumption.

Peptides contrasted these outcomes by utilizing seeming to be emissaries, which flag skin cells to deliver additional collagen while supporting mended tissues. This biological signaling re-establishes the youthful appearance and complements the resilience of the pores and skin to environmental insults. Peptides can be multi-functional and used for many unique skin needs and conditions.

Anti-Aging Peptides, The Powerful Skin Repairers

Demand for anti-aging peptides has grown extensively in the beauty industry due to their effectiveness in fighting the visible manifestations of aging. These peptides can be grouped based on their movement mechanisms, such as signal, carrier, and enzyme-inhibitor peptides.

Signal peptides (Palmitoyl pentapeptide—better recognized as matrixyl): They have an identical feature to mild retinoids; however, they work via stimulating fatty fibroblasts to boost collagen and elastin manufacturing. Clinical studies have demonstrated that matrixyl can decrease wrinkle intensity and improve skin elasticity, making matrixyl a key ingredient in anti-aging formulations. Acetyl hexapeptide is another widely used signaling tech called "Botox in a jar," and that can cause the facial muscular tissues to loosen up and save you the looks of expression traces.

Enzyme-inhibitor peptides act by inhibiting the activity of Collagenase enzymes (MMPs), which are thought to degrade collagen and elastin. Enzyme-inhibitor peptides help prolong the life span of those structural proteins, which helps maintain pores and skin integrity and delay the appearance of aging.

The combined effect of those anti-aging peptides is a youthful-looking, radiant skin. Since they may be capable of turning on, at a minimum, a few paths of the aging process, they are mighty in skin care regimens that target wrinkles, grow skin firmness, and promote universal skin health.

Skin-reforming hydrating peptides

Peptides are also used in their anti-aging houses and play an essential role in skin regeneration and hydration. Skin regeneration involves exchanging old, damaged cellular elements with new, healthy, functioning cells to maintain skin strength and heal wounds. Peptides such as hexapeptides and tetrapeptides promote wound healing by enhancing cell proliferation and migration. They also trigger the production of expansion components that encourage tissue recovery and renewal.

The different side of skincare with hydration is that moisture levels immediately influence skin elasticity, texture, and overall appearance. Dry skin often looks lackluster, coarse, and sensitive. Hyaluronic acid and aquaporin peptides are examples of peptides that address this problem by enhancing the skin's ability to retain moisture. Hyaluronic acid peptides are vital to improve hyaluronic acid, a haunt of water molecules, and receive additional humidity. On the other hand, aquaporin peptides facilitate water transportation inside the skin by modulating aquaporin channels, which becomes instrumental in maintaining optimal hydration levels across the different skin layers.

Not only do these hydrating peptides assist in restoring moisture, but they additionally strengthen the skin barrier, which is the primary line of protection from outside aggressors. A strong skin barrier helps to maintain moisture and protects against irritants, allergens, and pathogens. Combining this hydration and barrier reinforcement leads to smoother, plumper, healthier-looking skin.

How Peptides Can Be Used To Target Specific Skin Concerns

Peptides can help target specific skin concerns, from pigmentation and acne to sensitivity and redness, making them one of the most beneficial ingredients in cosmetics!

Hyperpigmentation Peptides

Hyperpigmentation: The basis of uneven pores and skin tone and hyperpigmentation is due to the extra manufacturing of melanin. Oligopeptides and decapeptides that block melanin biogenesis target melanogenesis, the process of melanin synthesis. They block the action of tyrosinase, a key enzyme in melanin production, and those peptides also help lighten dark spots and skin tone. They're gentle enough to use even on delicate pores and skin, which is especially nice for people who cannot tolerate conventional depigmenting brokers (ahem, hydroquinone).

Anti-Acne & Inflammation Peptides

Acne is another common skin issue characterized by inflammation, clogged pores, and bacterial proliferation. Besides defensins and cathelicidins, antimicrobial peptides (AMPs) are critical in combating acne, as they first hit acne-causing bacteria such as Propionibacterium acnes. These peptides also display anti-inflammatory properties, allowing less redness and inflammation associated with acne lesions to develop. Their double movement is not the most compelling sound or most earnest existing breakouts but prevents fate events.

Gentle Hydrating Skin Peptides

Peptides are soothing, particularly for sensitive or reactive skin. Specific peptides (such as palmitoyl tripeptide-8) are acknowledged to reduce infection and strengthen the skin barrier. Those peptides reduce signs and symptoms of inflammation, including redness and soreness, by modulating the release of pro-inflammatory cytokines. These skin care products provide a mild but effective treatment for these conditions such as rosacea and eczema.

Applications of Peptides in the Cosmetic Industry

The field of beauty peptides continues to evolve, driven by advances in peptide synthesis, delivery systems, and ingredient science. Investigators are developing new peptides with improved stability, bioavailability, and activity. One of the difficulties with peptide skincare is ensuring that those molecules stay firm and active until they hit their target websites in the skin. This challenge has been partly solved through various encapsulation techniques, such as liposomes and nanoparticles, which can move defensins away from degradation and allow for their controlled release.

Another exciting development is that of biomimetic peptides that mimic the structure and function of peptides happening in the structure. This is where peptides can provide next-level compatibility with pores and skin body structure, and due to being bio-individual, amalgamated to deal with the specific pores and skin needs. For example, biomimetic peptides that imitate the movement of herbal growth factors are being developed to improve skin regeneration and anti-aging effects.

In addition, the increasing need for natural and clean beauty products has driven the enhancement of beauty peptides. The research now is on the production of peptides based on renewable resources and their application in green formulations. This technique keeps the patron's selections for effective and promising products to the surroundings.

Will consumers embrace the new design or find it off-putting?

Incorporating peptides in beauty products has developed skincare technology and changed purchaser expectations. Today's shoppers are more intelligent and cautious, looking for products backed by medical research and clinical evidence. Their proven effectiveness and safety have given Peptides a significant reputation in high-end and mass-market skin care products.

Cosmetics based totally on peptides appear to have a sturdy boom charge because of the bigger cognizance of the peptides' blessings and the growing call for anti-getting old and multifunctional beauty skin care solutions. As per various industry reports, the global cosmetic peptides market is still growing, driven by enhancement in the Peptide era and increasing client preferences through cosmetic peptides.

The offer of focused, tailored solutions for various skin problems. Peptides have transformed the cosmetics sector. These bioactive molecules have derived the fast and have laid the bar when it comes to growth for skin care, from getting old and hydration to pigmentation and zits. They are critical players in searching for younger, glowing pores and skin because of their ability to modulate cellular processes and decorate pores and skin health. With the steady increase of studies and innovation in peptide technology, the destiny of beauty peptides seems to be even more promising — meeting the ever-evolving desires of customers and shaping the subsequent technology of skin-care merchandise

4.3 Diagnostic and Research Peptides

Peptides are also precious in state-of-the-art diagnostics and biomedical research because they uniquely bind to various organic targets. Due to their relatively small length, synthetic nature, and inherent flexibility, they can have high specificity and affinity interactions with proteins, nucleic acids, and other biomolecules. Such attributes make peptides indispensable for disorder detection, monitoring, and research, where high sensitivity and specificity are paramount. Diagnostic and research applications of peptides have allowed us unprecedented means of managing complex diseases and opened an exciting frontier in developing the next generation of personalized medicine.

Peptides in Molecular Imaging

Molecular imaging, where advanced imaging technologies are adapted to study cellular and molecular approaches in intact organisms, was only in its infancy and is now pretty advanced. This proximity has gained prominence because peptides can be specifically targeted by dealers capable of selectively binding to particular sickness markers. Such capacity to target specific molecular signatures in pathological conditions like cancers or infectious diseases makes them ideal candidates for imaging applications.

In particular, radiolabeled peptides have proven to be exceptional tools for molecular imaging. Radiolabeling peptides using radioactive isotopes can result in effective diagnostic marketing in positron emission tomography (PET) and unmarried-photon emission computed tomography (SPECT). These imaging modalities provide unique, 3D snapshots of biological behaviors in real-time, delivering exceptional information on the location and behavior of disease sites.

The most extensive programs in radiolabeled peptides are in cancer diagnosis and monitoring. Peptide-based imaging: Tumor-associated antigens (i.e., proteins or other molecules present at or near the antigen). For instance, PET imaging with peptides targeting either somatostatin receptors, frequently found overexpressed on neuroendocrine tumors, has been successfully employed. Radiolabeled peptides bind to those receptors, allowing clinicians to accurately localize the tumor, assess tumor load, and determine treatment response.

Outside of oncology, peptide-based totally for a more extended period molecular imaging has visible applications in cardiology, neurology, and infectious illnesses. In cardiovascular medications, radiolabeled peptides might be used to eliminate atherosclerotic plaques and examine myocardial perfusion, which is helpful for early analysis and control of heart disease. For example, in neurology, peptides enriched on amyloid-beta plaques and tau proteins, which are the pathological features of Alzheimer's disease, have been developed as imaging agents to track the progression of neurodegenerative diseases. Likewise, in a tactic of contagious disease, peptide-based imaging agents will enable the detection of infection or infection sites and supply treasured facts for treatment-making plans.

Peptides supply excellent promise for molecular imaging and a growing field of theranostics that combines therapeutic and diagnostic abilities in a single agent. Peptides utilized in theranostic packages can provide therapeutic retailers directly to illness sites at the same time as at the identical time, permitting actual-time tracking of remedy efficacy. This dual functionality is the primary leap ahead in personalized medicine and presents sufferers with personalized and compelling therapeutic strategies.

Identification of Early Biomarkers and Disease Monitoring

Biomarkers are biological indicators that reflect a disease's presence, progression, or severity. Identifying and measuring biomarkers are critical to disease prognosis, characterization, and surveillance. Due to their ability to specifically bind to target molecules and compatibility with different detection systems, Peptides have become powerful tools in biomarker discovery and analysis.

Peptide-based total assays were fashioned to identify several disease-unique biomarkers. For instance, in cancer diagnostics, peptides that target proteins such as HER2, EGFR, and PSA (prostate-unique antigen) are broadly used for identifying and categorizing rare types of tumors. They allow early diagnosis, help determine the severity of the disease, and assist in treatment decisions.

Peptides are essential in cardiovascular studies to isolate coronary heart sickness biomarker regulators, troponin, and natriuretic peptides. Higher levels of those biomarkers represent myocardial injury or heart failure, providing valuable information for early intervention. Likewise, within the subject of neurodegenerative diseases, peptides are utilized to discover such biomarkers as amyloid-beta and tau proteins, critical players in Alzheimer's disease and other forms of dementia.

Exploiting this versatility, multiplexed assays can be developed based on peptides' ability to detect multiple biomarkers simultaneously. This ability is particularly valued in complex diseases with numerous pathways and molecular alterations. Peptide-primarily based multiplex assays grant a comprehensive image of a patient's biomarker profile, enabling correct characterization and subgrouping of patients according to their sickness characteristics. Importantly, this stratification is vital for applying for precision medicine, where treatments are customized based on the individual needs of tumor patients.

An exciting area of research is the design of peptide-based biosensors. The gadgets use peptide and advanced sensing technology, such as electrochemical or optical detection, to enable relatively sensitive and portable diagnostics. Peptide-primarily based biosensors have been utilized in factor-of-care assessment, allowing speedy and accurate analysis of disorders similar to infections, metabolic dysfunctions, and cancers. They are especially treasured in asset-restrained settings, wherein the right of entry to conventional diagnostic amenities can be limited due to their ease of use and fee effectiveness.

Innovation in Precision Medicine

Personalized medication or precision medicinal drugs pursue the goal of customizing scientific treatment to the individual man or woman affected person. Peptide-primarily based tools enable the identification of specific molecular signatures associated with a given disorder in a patient and have driven the development of this diagnostic strategy. That information can also be used to select the most effective treatment, track treatment response, and adjust interventions accordingly.

Peptide-primarily based diagnostics can enable a real-time readout of an affected person`s standing, making this one essential benefit to any customized medication option. Peptide-based imaging and biomarker assays quantifying the response to chemotherapy or targeted therapies in most cancer remedies allow clinicians to make data-driven choices about persevering with, titrating, or halting remedies. Along with the potential for attaining great results, it also facilitates a comment loop that will reduce the risk of bad results.

Peptide-based total diagnostics no longer solely help in informing the options of the remedy but additionally in predicting the disorder's danger and development. Peptide-primarily based tools can stop the manifestation of pleasant conditions by subsequent identification of biomarkers related to a primary-stage ailment or predisposal genetic elements to pinpoint individuals in excessive danger of developing high-quality situations. This predictive ability enables action (before the disease appears, including lifestyle changes or even preventive treatment plans), thus minimizing the disease and improving long-term health outcomes.

Furthermore, peptide-based diagnostics can be incorporated into tailored medicinal drug workflows, further supported by differences in bioinformatics and machine studying. This technology facilitates the analysis of large-scale datasets produced by peptide-based assays, uncovering patterns and relationships that may be difficult to detect through traditional analysis methods. Exploiting these insights enables researchers and clinicians to create better models for prediction and hone their understanding of disease processes.

The Use of Peptides in Research and Drug Development

Peptides are not the most effective and beneficial in diagnosis, but they are also valuable and obligatory for biomedical research and drug advancement. This capacity to modulate organic pathways and interact with specific targets makes these essential tools for investigating disease mechanisms and attractive therapeutic targets.

Peptides are also applied in research and are often used as probes to study PPIs (protein-protein interactions), enzyme sports, and mobile signaling pathways. Fluorescence-labeled peptides, for instance, can be applied to observe the dynamics of cell events in real-time and provide new information to understand the mechanisms of disease development. Peptide inhibitors or agonists can be delivered in a manner that allows the investigation of the role of unique pathways in health and disease similarly.

These are also employed in usher screening assays for drug discovery using peptides. These assays include screening large libraries of compounds, including peptide libraries, to identify molecules that exhibit specific biological activities. Once a viable candidate has been identified, it could be refined and developed into a therapeutic drug. This is where peptides have so many benefits in this context, as well as their ability to target complex and previously their targets that were not amenable to drugs, such as protein-protein interactions.

In addition, advances in peptide engineering and delivery technologies create a windfall for therapeutics. To get around the limitations of typical peptide capsules, peptides structurally altered towards a more favorable balance of bioavailability and objective specificity are being brought through the various phases of advancement. These improvements are expanding to select sicknesses that can be correctly handled with peptide-based therapies. These include instances that had been regularly tough to handle with small-molecule or biological pills.

Outlook and limitations

Peptide-primarily based diagnostics and studies continue to evolve, stimulated through cellular advancements in peptide chemistry, biotechnology, and statistics evaluation. Microfluidics, nanotechnology, and emerging technologies are expected to further enhance the capabilities of peptide-based diagnostic tools, leading the way for susceptible and miniaturized devices for point-of-care testing.

But we still have a few challenges to solve. The synthesis and production of peptides may be complex and costly, especially for large-scale packages. To ensure high adoption of peptide-based total diagnostics, there is a need to streamline production approaches and pressure fees down. Moreover, peptide-based products must be carefully manipulated to ensure their stability and shelf lives, preventing incapacity or inconsistency.

This is also of crucial importance when it comes to regulatory and scientific validation. Peptide-based diagnostics require exceptional validation for protection, precision, and medical use. They will have the power to carry novel peptide-based alternatives to the market if we navigate those challenges collaboratively with researchers, business stakeholders, and regulatory groups.

4.4 Biotechnological and agricultural use of peptides

The significant possibilities of peptides are far past their deals in fitness and diagnosis and find massive use in biotechnology and agriculture. Peptides play an essential role in those industries as a part of cutting-edge solutions that address many of the most critical global issues, such as food security, sustainable agriculture, and livestock health and wellness. With populations rising and natural resources under increasing threat, the multitude of developments in agriculture will be essential in providing a system capable of producing high levels of food production in an efficient, sustainable, and safe manner, now more than ever. The flexible biochemical properties of peptides have them uniquely positioned to meet those needs.

Use of peptides in crop protection and growth promotion

Proper sustainable crop protection is one of the tasks of agriculture innovation. Chemical crop protection has been the backbone of agricultural practices for centuries. However, these strategies have serious disadvantages, including potential environmental contaminants, non-target harm, and resistance development in pests and pathogens. Additionally, peptides represent a hope for the future through the generation of bio-insecticides and bio-stimulants.

Peptide-based bio-insecticides are aimed at selective pest or pathogen attacks and do not harm beneficial organisms or the environment. Peptides have been used for various things, but these peptide-primarily based retailers just function by disrupting essential organic techniques within the pest, comprising enzymatic capabilities or mobile integrity. Through contact, for instance, positively-charged peptides (AMPs) may pass through microbial cellular membranes, leading to leakage of cytoplasmic/mobile contents and, consequently, cell dies. In the case of crop disease, this mechanism of action is especially coveted for disease prevention of fungal and bacterial pathogens.

Apart from pest manipulation, they are vital for boosting the increase and resilience of the crops. Peptide-based bio-stimulants have been shown to enhance metabolism apparitional nutrient uptake and augment growth resistance to abiotic stresses such as drought, salinity, and high automatic temperatures. These bio-stimulants are paintings when you consider that they modulate plant hormone levels and activate pressure-responsive pathways that enhance boom performance under demanding environmental conditions.

Peptide-based crop safety and growth enhancement also have a significant environmental advantage. Peptides are biodegradable and do not leave toxic residues in the soil or water, unlike synthetic chemical compounds. This is why they are often considered the ideal plants for sustainable agricultural production, seeking to minimize the environmental impact of crop production. The specificity of completely bio-insecticide-based peptides reduces the chance of resistance evolution in pests, ensuring a lengthy period.

For Animal Health and Welfare Applications:

Domesticated animals and aquaculture are the backbone of the international meals system, providing wealthy protein meals for human nutrition. However, these animals are always at risk of diseases and infections, affecting their health and proper functioning. Historically, the principal strategy of disorder prevention and remedy has been through the overuse of antibiotics in animal husbandry. Nevertheless, this exercise has led to the rise of antimicrobial-resistant pathogens that risk animal and human health. Peptides possess an alternative as antimicrobial vendors since they can combat infections without conferring antibiotic resistance.

Antimicrobial peptides (AMPs) are rushing in-place molecules are vital in introducing protection from pathogens in several organisms. They display a broad-spectrum interest against bacteria, fungi, and viruses these peptides. In farm animals and aquaculture, AMP use can effectively prevent and treat infections, as well as reduce the use of traditional antibiotics. Previously, AMPs comprising defensins and cathelicidins have been reported to exhibit potent antibacterial activity in hens and fish.

Aside from their antimicrobial properties, Peptides are also critical for improving the health and productivity of an animal. Some peptides are growth promoters that help in muscle development and enhance the feed conversion ratio. Peptide-based total growth factors can, for example, improve protein synthesis and increase the rate of weight gain in cattle, resulting in higher meat products. Peptides were used in aquaculture to enhance the growth and survival prices of fish and shellfish, thereby supplying a greater dependable and sustainable supply of food.

Peptides in livestock are also linked to animal immune modulation, which helps cattle to deal with infection and disease recovery more adequately. Peptide-primarily based goods can increase pets' natural defense mechanisms by optimizing the immune cells' pastime and merchandising the manufacturing of antibodies. Not only does this not reduce sickness, but it also diminishes the want for clinical involvement, resulting in more excellent, affordable veterinary care and higher animal welfare.

Using peptides signifies a new hope in vertical farming and sustainable agriculture.

The worldwide agriculture industry is under monstrous stress to grow sustenance manufacturing while minimizing its environmental effect. Peptides provide new solutions that complement sustainable agriculture standards, aiming to stabilize output with ecological stewardship. Peptides reduce dependence on agrochemical inputs and optimize the operationality of the herb by rendering effective and sustainable farming systems.

Handling soil fitness, the basis of long-term productiveness in agriculture is one of the most demanding situations. Peptides contribute to soil fitness by stimulating the production of beneficial microorganisms that improve soil fertility and structure. For example, peptides can help trigger the activity of nitrogen-fixing bacteria that turn atmospheric nitrogen into a form that plants can use. It minimizes the requirement of synthetic nitrogen fertilizers, which are power-hungry to create and can result in environmental problems such as water pollution and greenhouse fuel emissions.

Peptides also help to reinforce the efficiency of nutrient usage in plants. Peptide-inspected bio-stimulants expand the absorption and utilization of essential nutrients (including phosphorus and potassium) and reduce fertilizer intake for top crop performance. Not only does this minimize production expenses, but it also profits the surroundings by conserving agricultural practices.

Water sources play a second indispensable notable sustainable agriculture question. Developments in peptide-primarily based products may increase the water-use efficiency of crops to retain increase and productivity under water-limited conditions. This is particularly precious in regions with excessive aridity and semi-aridity where water scarcity presents a giant project to agricultural production.

After all, as they say, the only thing predictable about the telecommunications vertical is that it will always be unpredictable.

Using peptides in biotechnological and agricultural applications is a giant step in the direction of the demanding situations of meal safety and sustainability.

However, some challenges can prevent the extensive adoption of peptide-based products. Several boundaries exist, the main one being the production cost, which might be lower than that of conventional agricultural inputs. Improvements in peptide synthesis and production technologies are essential to lower costs and increase the availability of these products for farmers.

Regulatory environment: may differ significantly across countries. The approval process for new peptide-based products can be intricate and prolonged, potentially hindering their time-to-market. This is important for the market access, development, and regulatory submission of peptide-based total innovations, and consequently, clean and regular regulatory frameworks must be implemented to help them.

Nonetheless, notwithstanding these tough ought-to-faces, the prospects for peptides for biotechnology and agriculture are extraordinarily bright. Insights into novel functions and mechanisms of action for peptides continue to emerge, expanding their potential impact. Peptide-primarily based vaccines, which have the gain of simply activating immune responses that may be then implemented as secure and robust merchandise without the inducement of immunity, may want to lead to the transformation of illness manipulation and prevention methods in regions of splendid financial significance, such as cattle and aquaculture.

Also, genetic engineering and artificial biology developments open up new venues for peptide production and optimization. This technology makes it possible to design peptides with better characteristics, such as increased stability, selectivity, and activity. Such ought to provoke the advance of more potent and extra versatile 2nd technology peptide-based products.

Peptides have become powerful tools in biotechnology and agriculture, providing innovative solutions to some of our most challenging issues. Their versatility and power to operate these organic and green agriculture practices offer them additional crop protection, boom-loading, and animal comfort choices. However, with research and technology set to improve, peptides may become more integral to agriculture and biotech's future. The unique residences of peptides can be utilized to produce farming systems that are efficient enough to scale up but also ecologically sustainable to guarantee the security and sustainability of a meal delivered for the next sustainable generations.

Chapter 5: Peptide Design and Scoring

Due to their excessive specificity, low toxicity, and diverse biological sports, peptides have emerged as a widespread class of molecules in each therapeutic and diagnostic program. The demand for simple peptide sequences has increased as research continues to find the multifaceted roles that peptides perform in cell features. Peptides offer great potential for drug development, biomarkers, discovery, or novel fabrics, but designing and optimizing peptide sequences is critical for us to exploit such capacity. This bankruptcy addresses the ideas and practices that inform peptide series layout: the gear, techniques, and actual global implementations that are some cornerstones of this dynamic field.

5.1 Basics of Sequence Design

The design of peptide sequences is a sophisticated process that demands complete knowledge of the molecular characteristics dictating their structure and activity. Peptides, polymers of amino acid residues connected by peptide bonds, are versatile organic molecules that can be used for diverse functions, ranging from enzymatic activity and signal transduction to immunomodulatory and therapeutic interventions. Their biological activity is directly associated with their series of 3-dimensional conformation, stability, and their association with organic objectives. Thus, peptide design is a multi-faceted effort to achieve a compromise between potency, strength, and bioavailability.

Peptide design is gaining more attention recently, especially in developing peptide-based drugs, materials, and diagnostics. As we face increasingly complex biological problems, it will be essential to be able to design peptides that can execute specific functions effectively. This involves selecting the correct amino acid sequences and providing structural tasks to the peptide, making it work better in its respective environment. Whether the objective is a therapeutic agent, a biomarker, or an ultimate fabric, the principles of peptide sequence design continue to be the underpinning of success.

A few Aspects of Stability and Activity of Peptides

The cornerstone of functional overall performance is peptide balance/efficacy, and expertise in those parameters is critical to guiding robust design. Stability is a crucial indicator for a peptide to resist degradation under different conditions, including exposure to proteolytic enzymes, pH change, and temperature variation. Efficacy, on the other hand, relates to the ability of the peptide to bind to its target precisely and to induce the correct biochemical response.

However, a central dilemma is the protection of peptides from degradation by proteolytic enzymes, which are plentiful in both the extracellular and intracellular environments. Ultimately, proteolysis offers to diminish the energetic lifespan of peptides, but it also adds to their immersion utility. Approaches containing non-herbal amino acids, cyclization, and enhancing terminal residues have been employed to beautify proteolytic stability.

Another important factor regarding the design of peptide series is the balance between hydrophobicity and hydrophilicity. Typically, hydrophobic amino acid residues constitute a large part of the structure core of the peptide; stability is derived from intramolecular interactions (hydrogen bonding and van der Waals). These interactions are significant for maintaining the secondary and tertiary structures: alpha-helices, beta-sheets, and beta-turns. In contrast, hydrophilic residues can play an essential role in both solubility and interaction with the target. They are usually disclosed on the peptide's floor, exposing them to interactions with aqueous environments and polar practical corporations of biological targets.

The secondary and tertiary structures of peptides profoundly affect their relative stability and biological activity. For example, intra-helical hydrogen bonding stabilizes alpha-helices by maintaining a small, functional structure. On the other hand, beta-sheets depend on inter-strand hydrogen bonds, which may affect the peptide's stress and stability. This means these structural motifs typically act as recognition factors in receptor binding and are essential for the peptide's functionality.

In addition to structural stability, other physicochemical properties, including charge distribution, molecular weight, and pKa, affect peptide function. The presence of positively or negatively charged peptides can make the peptide interaction with cellular membrane interfaces or protein interfaces favorable or unfavorable by interaction with the negatively charged. The sequence design in fine-tuning those homes in a particular ecological context could ensure the most productive overall performance.

Design Strategies for Typical Sequences

Peptide collection is an adaptive process that utilizes different strategies to achieve high performance and adapt to specific hurdles. These tactics regularly combine empirical expertise, experimental validation, and computational insights to achieve desired outcomes.

Integration of Unnatural Amino Acids and Peptide Mimetics

Adding unnatural amino acids or peptide mimetics enhances peptide balance and function. These non-canonical building blocks mimic natural amino acids' structural and beneficial properties while providing additional benefits. Drug-like peptides can be obtained, for example, by incorporating beta-amino acids or N-methylated residues, which increase proteolysis resistance by modifying the peptide backbone and preventing efficient proteolytic cleavage. Likewise, peptoids (N-substituted glycines) provide structural diversity and higher stability.

Cyclization is a broadly used alternative to assist with balance and tension. Cyclic peptides are also far less susceptible to enzymatic degradation because of a covalent bond formed between the N-terminus- and C-terminus or between facet chains of particular residues. Cyclization can also enhance binding affinity by pre-organizing the peptide in a biologically active conformation and lowering the entropic loss upon complex formation with a target.

The enantiomers of the genetically encoded natural L-amino acids, D-amino acids, are often incorporated into peptide sequences to enhance stability.

Although most proteolytic enzymes are stereospecific and recognize only L-amino acids, incorporating D-amino acids can substantially improve a peptide's resistance toward enzymatic degradation. This method is precious when designing peptides intended for prolonged systemic transport.

Truncation of the sequence and optimizing

Regular strategies involve truncating peptide sequences optimizing to keep or decorate activity while reducing the charge of synthesis and potential side effects. This strategy has the utmost problem of defining the minimal energetic series, the shortest peptide fragment that preserves total organic activity. It streamlines the synthesis process and minimizes the potential for non-specific interactions and immunogenicity.

Alanine scanning and site-directed mutagenesis are commonly used methods for identifying critical residues in a panel of peptides. Alanine scanning involves substituting each amino acid in the sequence with alanine and examining the impact on activity. This helps uncover residues that are fundamental for characteristic and can direct additional optimization work. In a golden standard related, website online-directed mutagenesis permits the focused alteration of specific residues, offering records on the contribution of those residues to structural stability and binding to the target.

Greedy Hydrophobic Moment Optimisation

Peptides often demonstrate amphipathic properties, containing hydrophilic residues on one side and hydrophobic residues on the other in their helical structure. This second hydrophobic region is responsible for membrane interaction and protein binding. Depending on the intended organic objective, peptides could be engineered with hydrophobic moments tailored to interact with lipid bilayers or to fill hydrophobic cavities in proteins.

Molecular Modeling and Computational Design

Computational gear contributes significantly to peptide design because it permits the prediction and optimization of peptide structure and functions. Molecular modeling, including molecular dynamics simulation and docking studies, provides valuable information regarding the conformational behavior of peptides and their binding to organic targets.

Under diverse factors, researchers have discovered conformational flexibility and balance-peptide in molecular dynamics simulations. These simulations find stable secondary and tertiary structures, predict folding pathways, and estimate the effect of various mutations in the series. Similarly, docking studies indicate the binding orientation and affinity of the peptides for their target molecules, which essentially guide the engineering of sequences with better binding properties.

Again, implementing computational libraries and collection databases performs due diligence within promising peptide candidate identifications. Through criticism of sequences with identified purposeful peptides, new peptides with similar or greater residences can be designed. Machine learning and artificial intelligence have additionally permitted the introduction of predictive fashions that examine peptide balance, solubility, and goal specificity from sequence records.

The Rational vs. Combinatorial Design

Peptide manipulation methods could be broadly categorized into move-forward manipulation and combinatorial concepts. Rational layout involves employing facts of peptide shape, target interactions, and physical/chemical mechanisms to synthesize sequences with prescribed houses. It is an exact method and allows the investigative control of peptide properties.

On the other hand, combinatorial approaches are concerned with generating large libraries of peptide sequences and screening these for some preferred traits of interest. Methods such as phage display and rest mRNA display are widely used to identify high-affinity binders from these libraries. While they are time-consuming due to the combinatorial processes involved, they have the advantage of uncovering sequences you would never have predicted by rational design.

PTMs: Post-Translational Modifications

Including publish-translational adjustments (PTMs) at some point of peptide sequences can beautify their balance and function through several orders of magnitude. PTMs such as phosphorylation, glycosylation, and acetylation may influence the peptide's solubility, stability, and interactions with its target. For example, phosphorylation of serine, threonine, or tyrosine residues can modulate the pastime of the infused peptide via modifying its interaction with dedicated signaling pathways.

Self-Assembling Peptides

The design of self-assembling peptides has attracted recent interest due to their potential utility in nanostructure and biomaterial creation. When an exact environmental stimulus is applied, these are prepared to spontaneously mold into orders of nanofibers, hydrogels, or vesicles. Beyond traditional therapeutic applications, the versatility of the peptide design is also evidenced by the presence of self-assembling peptides designed for drug transport, tissue engineering, and biosensing.

5.2 Computational tools for peptide design

The addition of computational tools has brought a paradigm shift to peptide design, changing it from a mainly experimental and lengthy process to both efficient and remarkably predictive. Computational techniques allow researchers to simulate the behavior of peptides in silico, facilitating the layout process and presenting new understanding that may be challenging or impossible to obtain using trendy experimental techniques. This section reviews the landscape of software programs and equipment for peptide layout (preferences, programs, and how they are helpful resources in peptide sequence fashion, optimization, and evaluation).

A Review of Peptide Design Programs

Peptide layout packages are computational gear that offers a broad spectrum of functionalities in predicting peptide systems, comparing peptide-goal interactions, and simulating peptide dynamics. These programs may be sought in lots of paperwork, from general-cause molecular modeling software programs to additional specialized devices focusing on precise features of peptide layout, including sequence prediction or folding. This post presents some of the most popular and best peptide design tools you can find for researchers today.

PyMOL: A Comprehensive Tool for Structural Visualization and Analysis

PyMOL is one of the most widely used tools for molecular visualization within the peptide and protein community. Although often referred to as a visualization tool, PyMOL can perform fundamental analysis and manipulation of protein and peptide structures. It can show peptide structures in many forms, e.g., ribbon diagrams, surface models, and space-filling fashions. This capability enables researchers to extract deep insights into peptide conformations while also being alerted to possible structural motifs of interest like alpha-helices, beta-strands, or loops.

Besides visualization, PyMOL is used to investigate molecular properties such as hydrogen bonding, hydrophobic interactions, and electrostatic potentials. They are valuable for expertise in how peptides interact with their targets, be it enzymes, receptors, or other biomolecules. The ability of PyMOL to be incorporated with other computational tools, such as docking programs and molecular dynamics simulators, also gives opportunities for peptide design and refinements.

Rosetta: De Novo Peptide Design and Folding Simulations

Introduction: Rosetta is a general-purpose software suite for protein and peptide design. Notably, it is recognized for its ability to execute de novo peptide design, where the endogenous sequence of a peptide is optimized from scratch based chiefly on desired properties such as stability, binding affinity, and biological performance. However, in contrast to more traditional experimental methods that include precise rounds of synthesis and testing, Rosetta allows researchers to execute hundreds of simulations in a fraction of the time, testing different peptide sequences and straightforwardly fooling styles.

Backbone Conformation Prediction: Rosetta uses an advanced strength function to predict the best conformations of peptide sequences. This characteristic analyzes the collectivity of elements, such as van der Waals, hydrogen bonds, and solvation electricity, to discover the lowest strength doses of the peptide. By sampling those conformations, Rosetta can identify sequences most likely to fold into a relatively stable, functional structure or bind with high affinity to a specific target molecule.

Rosetta is also quite powerful for shape prediction (the computational technique that uses physical movement to predict the shape of a 3D object such as proteins), protein-peptide docking simulations, and refinement of peptide-based drug candidates aside from de novo design. The software program has been used successfully in many applications, including designing peptide inhibitors of protein-protein interactions and optimizing peptides for nanomaterials.

Schrödinger Suite: Complete Drug Design

This package includes a deed of integration providing extensive spectrum peptide design functionality, making it one of the foremost comprehensive day program tools. Another central platform, Schrödinger, has made their specific applications available here: molecular dynamics, protein-ligand docking, and peptide folding predictions. Pepscan is commonly utilized in drug discovery and material science, providing valuable records on peptide conduct and binding with organic targets.

Desmond, a molecular dynamics device from Schrödinger, allows us to simulate the motion of the atoms in a peptide structure over the years, giving some information about the stability, flexibility, and behavior of a peptide in solution. Maestro interface contained in the Schrödinger suite also includes peptide modeling tools, enabling customers to expect secondary and tertiary peptide systems, layout peptide libraries, and carry out collection alignments.

The docking studies with the peptides against their target proteins are performed using Schrödinger's AutoDock and Glide modules. Such docking simulations are essential for knowing how well and how exclusively a peptide will bind to its target, helping to orient the blueprint of peptides with enhanced therapeutic potential.

PeptideBuilder and PEP-FOLD: De Novo Peptide Design and Folding Prediction Tools

PeptideBuilder and PEP-FOLD are tools geared explicitly towards researchers in the layout of peptide sequences de novo and their prediction, respectively. Such packages are especially useful in generating new peptide sequences and evaluating the likelihood of forming stable biological structures of relevance.

PeptideBuilder is a web application that quickly enables the pseudo definition of peptide sequences. It allows researchers to input a set of preferred peptides and instantly see the corresponding structure. It can also be used as a framework for designing peptide libraries by creating diverse sequences based on a consumer-defined template. PeptideBuilder can predict primary structural features such as alpha helices or beta strands, providing invaluable preliminary insight for more tailored design.

Nevertheless, PEP-FOLD is focused on correctly predicting peptide folding. As an extension, PEP-FOLD computes the corresponding predicted 3D conformation of the peptide at the enter collection and describes the probably secondary and tertiary structure. This is fundamental for specialization, the support of the peptide, its connection with other particles, and its general properties. It uses an energy minimization technique to refine its predictions, one of the most refined machinery available for de novo studio of peptide folding (PEP-FOLD).

Many different docking programs are available for docking small ligands to targets, such as AutoDock and ClusPro for docking peptide-target interactions.

The interplay between peptides and their targets is an aspect of peptide design that is especially important for therapeutic applications. Docking packages with AutoDock and ClusPro are invaluable tools for this approach. Using these bundles, analysts can reproduce the tying of a peptide to a specific receptor, compound, or different natural targets, giving essential records on holding quality, collaboration destinations, and the instrument of the board.

AutoDock, one of the most popular docking programs, is an efficient and flexible solution. Or it can be docked for peptides to various forms of receptors or protein goals, forecasting the high-quality binding conformation primarily based totally on molecular docking simulations. The ability of AutoDock to perform virtual screening on large peptide libraries makes it a powerful tool for peptide screening, especially in drug discovery.

ClusPro is another docking tool that is quite efficient in modeling interactions between peptides and proteins and between two proteins. It uses a clustering algorithm to predict the most likely binding modes of peptides to their targets, providing a set of candidate complexes that can be further tested experimentally. The speed at which ClusPro runs and the inconceivable finesse of the program could make ClusPro the perfect large-scale peptide layout software.

Advantages of In Silico Study

In silico assessment gives several advantages over experimental strategies, such as less expensive, time- and aid-depletion. This, through the clever use of computational tools, permits researchers to simulate the behavior of peptides without the need for costly and time-consuming laboratory testing. In silico approaches also allow for the high-throughput screening of large sequence spaces, allowing for the testing of hundreds or even thousands of peptide variants in a short period.

It saves time and money.

The most valuable advantage of in silico analysis is the reduction in the cost and time associated with building peptides. As traditionally practiced, peptide design would typically be limited to synthesizing a few peptide sequences, followed by organic testing of their stability and activity. This method is regularly the most costly and can take months or even years to provide excellent results.

However, in silico tools allow hundreds of peptide sequences to be arranged and tested in a few days or weeks. This ability to high-throughput profile peptide libraries dramatically accelerates the discovery of the most favorable candidates for laboratory evaluation and minimizes the cost of peptide development.

More Insight Into the Behavior Of Peptides

Computational tools also offer a more in-depth insight into peptide dynamics at the molecular level. Molecular dynamics simulations of interactions among peptides and their targets can yield valuable information about the specific residues that contribute to binding, the orientation of the peptide within the binding website, and the nature of the interactions that stabilize the complicated. This knowledge is essential for directing similar design cycles and improving peptide specificity and potency.

In addition, in silico analysis can be observed for hidden structural motifs or stabilizing interactions that are not yet obvious through experimental methods. For example, framework-primarily based part approaches can find transient conformations that account for a peptide's hobby, or docking research may even reveal diffuse adjustments in binding affinity that modulate overall net drug efficacy.

Unpacking Non-Obvious Design Solutions

A second advantage of in silico approaches is the capacity to push layout answers past the intuitive. Computational methods often propose peptides or structural patterns that traditional experimental methods might not have considered. An example might be a docking program that suggests a peptide sequence with a higher predicted binding affinity for an object receptor than expected solely from previous information regarding the receptor's structure.

The boons and intensity of this capacity to yield surprising responses is low in the form of new peptide-based therapeutics, inhibitors, and diagnostics that will now not be detected "by using methods" helper experiment and blunders approach. Hemolytic peptides are typically optimized in two stages: First, computational gear is used to model the outcomes of various changes to the peptide, enabling the researchers to pick sequences predicted to have better overall performance, specificity, and stability.

Incorporation of Artificial Intelligence and Machine Learning

The addition of generated intelligence (AI) and method learning (ML) has a superior peptide design. They also can analyze vast data sets to identify trends and patterns that might be difficult for humans to find. The type of peptide design offered by AI and ML algorithms is based on large libraries of known peptides that expect peptide residences, optimization of collection design, and automobile-precision predictivity of peptide activity.

Machine learning models are trained on large datasets of peptide-receptor interactions to predict which sequences are most likely to bind successfully to their targets. This allows for extensive specialized design, particularly for challenging applications such as developing particular peptides targeting complex proteins.

5 Studies of Peptide Design

Peptide design is a fast-moving field with far-reaching impact across various disciplines, particularly medicine, biotechnology, and materials science. This ability to create peptides with specially designed biological lifetimes has led to revolutionary progress in therapeutic and diagnostic development and new technologies. Here, we highlight some of the peptide-based packages with the relevant super case studies during this phase, with particular attention paid to the packages' success in pharmaceutical drug design and cutting-edge studies projects.

Getting Peptide-Based Drugs Into Hot Water

In the past few decades, peptides as healing agents have grown relatively noteworthy interest, presenting a compelling alternative to traditional minor molecule additives and biologics. Their relatively modest size, high specificity, and ability to engage complex biological devices render them invaluable for treating several diseases, from metabolic disorders to most cancers.

Peptide therapeutics are best characterized by insulin, a peptide hormone that regulates blood sugar ranges in diabetic patients. The discovery of insulin in the early twentieth century was a breakthrough in diabetes treatment. Still, the ultrashort 1/2-lifetime of the endogenous peptide hormone presented a challenge for sustaining long-term blood glucose control. Improvements in peptide sequence design led to the development of insulin analogs that modulated their pharmacokinetics based on changes to their amino acid sequences, enabling better therapeutic utility. Long-acting insulin analogs, for example, have been designed to create stable complexes of glargine and detemir that release the hormone gently into the circulation, thus minimizing the need for frequent injections and maintaining more stable blood glucose levels throughout the day.

Assessment In assessment, fast-acting insulin analogs pardner aspart and lispro were engineered with changes in his amino acid sequences that immortalize on circumstances and timely in time the peak. These modifications enable more control of blood sugar levels, namely post-prandial (after meal) while reducing the potential for hypoglycemia (low blood sugar levels). They have improved the quality of life of patients, reduced the complications of diabetes, and have become a cornerstone in global diabetes treatment.

The design of peptide-primarily based drugs has additionally impressed many different vital advancements, reminiscent of the invention of glucagon-like peptide-1 (GLP-1) receptor agonists, which have transformed the remedy of kind 2 diabetes and weight problems. GLP-1 is a hormone that has an essential role in blood sugar regulation by increasing insulin secretion and suppressing glucagon release, thereby lowering blood glucose levels. Nevertheless, GLP-1 has a brief half of its existence inside the circulation due to rapid inactivation via the enzyme dipeptidyl peptidase-4 (DPP-4). To address this issue, scientists have developed GLP-1 analogs that are DPP-4 resistant, which can lengthen life by 1/2- and improve therapeutic effects. Exenatide, liraglutide, and semaglutide are examples of GLP-1 receptor agonists. These tablets, however not the handiest, improve glycemic management in patients with diabetes; however, they also provide sizeable weight loss further to glycemic management, making it one of the essential powerful medicines for weight loss, a common co-morbidity in kind 2 diabetes. The triumph of GLP-1 receptor agonists illustrates the promise of delightful optimizing peptide gathering, as it gives you a sharp attraction for solving the rising global health pop-up causes of arenas and obesity.

The development of peptide-based total antibiotics, especially those directed against bacterial resistance mechanisms, demonstrates the potential of the peptide in combating some of the greatest threats to existing therapy. Antibiotic-resistant microorganisms have emerged as a primary public fitness problem, and inside the new antibiotics, new cyclic lipopeptide daptomycin treating Gram-herbal bacterial infection with the aid of the use of methicillin-resistant Staphylococcus aureus (MRSA). Daptomycin passively disrupts the integrity of the mobile membrane of bacteria, leading to mobile death. Daptomycin and other peptide antibiotics are a new tool in the fight against resistant pathogens. Daptomycin was designed to resist the peptide sequence. In contrast, various other peptide antibiotics have been designed to serve as a model of function, providing the potential for custom peptides that present a solid alternative to traditional antibiotics.

For many of those examples, the transformational strength of peptide-based whole drug design, wherein optimizing peptide sequences for stability, potency, and bioavailability outcomes in significant improvements in patient outcomes, is shown. By modifying the sequences of peptides, scientists have enhanced the pharmacokinetics and therapeutic utility of peptide-based medicines, opening up new possibilities for extra-effective remedy alternatives for various diseases.

Examples from Research

Outside the pharmaceutical trade, peptide structure likewise enjoys a critical role in enabling studies throughout many scientific fields. Among the most relevant peptide layout packages remains the advancement of cell-penetrating peptides (CPPs). CPPs, which are short, usually positively charged peptide sequences, can permeabilize cell membranes and deliver cargo such as drugs, nucleic acids, and proteins into cells, a key obstacle in drug delivery systems. Such clever peptides have gained massive attention due to their potential to eliminate the need for more invasive drug delivery methods and to overcome the challenges of delivering biological drugs that otherwise would not cross the cell membrane.

The discovery of the TAT peptide from the HIV transactivator of transcription protein was a central milestone in developing CPP. Depending on the cargo's size and nature, the TAT peptide can enter cells via the plasma membrane by direct penetration or endocytosis. Originally discovered more than two decades ago, several CPPs, in addition to penetration, MTS (mitochondrial-targeting sequence) peptides, and polyarginine-based mainly peptides, have been evolved and optimized for a wide variety of applications in gene therapy, vaccine delivery, and drug delivery. With the help of these peptides, nucleic acids like siRNAs and plasmids, unable to penetrate through the mobile membrane, were delivered to the focus of genetic expression. So, this provided new methods for different types of genetic problems.

Besides their role as drug carriers, CPPs have been recruited into therapeutic schemes, including cancer treatment. Using peptides that are especially appropriate to pass through tumor cells to deliver chemotherapeutic agents directly to tumorous tissues may limit viable cell destruction and decrease unwanted side effects. Besides, CPPs have been implemented to target oncogenic proteins within cancer cells, providing an attractive approach to kill cancer by inhibiting the abnormal signaling pathways that drive tumor growth and metastasis.

Peptide design in studies with a potential application includes the enhancement of amyloid-inhibiting peptides for neurodegenerative diseases such as Alzheimer's. In Alzheimer's disease, amyloid-beta peptide aggregation forms plaques that impair neuron performance and contribute to cognitive decline. For this reason, researchers have designed peptides primarily associated with amyloid-beta and aggregation inhibition, certainly preventing the formation of such toxic plaques. This peptide is typically specified through a mixture of competitive inhibition and molecular enhancement, binding to hot spots of the amyloid-beta peptide that promote aggregation. Though amyloid-focused peptides remain predominantly experimental, they signify a potential avenue for therapeutic agents to slow or stop the development of Alzheimer's disease.

Peptides are additionally being designed to focus on protein-protein interactions (PPIs), a class of molecular interactions that can be notoriously tough to concentrate on using standard small molecules or antibodies. PPIs are involved in many cellular processes, and altered interactions have been associated with a range of diseases, such as cancer, autoimmune diseases, and neurodegenerative disorders. Researchers are developing new therapeutic strategies by designing peptides capable of disrupting or modulating those interactions. Peptides such as those inhibiting the p53-MDM2 complex regulating the tumor suppressor p53 are studied as treatment methods in cancers that express impairment p53 regulation.

The ability to design peptides that target specific protein interactions is a significant advancement regarding the potential for brand-new medication improvement, specifically in areas where conventional medication improvement has been unable to deliver effective treatment. These peptides can be engineered with excessive specificity to associate with unique regions of proteins involved in disorder processes, imparting a particular therapeutic strategy with restricted off-target effects.

The Future of Peptide Design

The design of peptides spans a wide range of areas, from healing drug preparation to revolutionary study applications, as exemplified by the interpolated case studies. The world of peptide technology keeps developing, and there are several exciting features in the future. Overall, the specificity of the designable functions is improving as peptide design and in silico modeling capabilities continue to advance to optimize peptide accuracy and performance. Artificial intelligence (AI) and device studying integrated into peptide design workflows are predicted to expedite this procedure and allow quick-via-peptide candidates with the best promise.

Moreover, advancements in peptide libraries and excessive-throughput screening technology allow researchers to display massive peptides for distinctive organic activities. Combining these contemporary technologies promises to expand the availability of peptides for therapeutic and research applications, enabling the development of patient-specific medicine adapted to individual patient needs.

Peptide layout is a quickly advancing area of study, and it is manageable for transformative software in a tremendous range of contexts. Peptides offer exceptional avenues for innovation in medicinal drugs, from optimizing peptide medicine for the treatment of diabetes and obesity to newly developing therapeutic approaches against cancer and neurodegenerative diseases. Technological development

Chapter 6: Building Your Laboratory & Safety

6.1 Essential Lab Equipment

An appropriately outfitted laboratory is the bread and butter of any medical or industrial research facility. Laboratories are where life-changing experiments, research, and innovations are conducted, producing results that benefit multiple fields, including biochemistry, pharmacology, molecular biology, and environmental technology. Such operations primarily rely on this crucial lab piece of equipment, which they depend on for conducting the processes efficiently, accurately, and safely. Accurate results can be acquired with the correct laboratory equipment while keeping the risks to the workers and the environment as low as possible. In the specialized field of biochemical or chemical laboratories, specialized equipment, including peptide synthesizers, chromatography systems, and other analytical instruments, has become a standard staple.

All of that apparatus is made to meet specific needs in the lab and is essential to certain clinical procedures. With knowledge of this equipment's relevant importance and functioning, laboratory experts can handle intricate studies and development processes. In addition, all successful laboratory facilities consist of protective protocols, proper maintenance, and training on using such units. Here, we can investigate the epicenter lab framework we utilized for cutting-edge science and synthetic exploration, focusing on peptide synthesizers and chromatography systems and clarifying their central part in exploratory work processes.

Peptide Synthesizers

Peptide synthesizers are dedicated devices that enable the synthesis of peptides by strong-section peptide synthesis (SPPS). Peptides are short chains of amino acids, and they are essential in many biological processes. Therefore, they are crucial for pharmaceutical tablets, vaccines, diagnostics, and biochemical studies. Peptide synthesizers automate time-consuming steps of peptide synthesis. Researchers can quickly, accurately, and consistently produce peptides using peptide synthesizers.

SPPS is based on the simple idea of a molecule, which can be a primitive amino acid attached to a solid support resin or guided by the connected device, which is the "scaffold" that a peptide chain may be assembled on the device. In this manner, peptide synthesis is a single amino acid at a time process. This is often one of the most tightly regulated collections, and the chain can be developed by enabling the peptide synthesizer. The process is chemistrically rigorous, involving iterative cycles of coupling (adding a new amino acid to the chain), washing (removing excess reagents), and deprotecting (removing protective groups on amino acids to allow further coupling).

Producing peptides for applications ranging from drug discovery through immunology and vaccine development to therapeutic protein engineering entails peptide synthesis as a critical component. The equipment used for peptide synthesis also proves invaluable in streamlining and facilitating the system. Peptide synthesizers are utterly automatic and have specific control over other response parameters such as temperature, pH, and concentration of reagents, resulting in high-precision peptide synthesis with high yield and purity.

Today, all peptide synthesizers have extreme levels of computerization, which helps minimize human errors and increase the efficiency of the synthesis method. These machines are usually established with pre-programmed protocols that control the critical chemical reactions in the peptide synthesis process, thus making it faster and more reproducible. The synthesizers can make peptides of different lengths, ranging from quick and straightforward peptides to low precision, but long, complex chains and broad precision and expertise are needed.

Peptide synthesis automation has made it easier to reproduce and save time for manufacturing by lowering exertion prices. Moreover, recent advances in the generation of peptide synthesizers, in conjunction with microwave-assisted peptide synthesis (MPS), have even increased the multiplicative of the technique by using microwave power to accelerate reaction rates. This generation has been confirmed to considerably lessen the time essential to synthesize peptides while retaining the purity and excellence of the final product. This is especially important when large sections of peptides are being synthesized for use in research, medical trials, or commercial production.

In microwave-assisted peptide synthesis, the reaction mixture is subjected to microwave radiation, which enhances the energy within the apparatus and further accelerates the chemical reactions. This allows faster coupling reactions, enabling researchers to reduce the overall time for synthesis. The microwave-assisted peptide was motivated as an appealing strategy for excessive-throughput peptide synthesis in research and industrial laboratories due to the acceleration of velocity and performance.

Peptide synthesizers are used not only to develop and improve synthesis speed but also to increase peptide purity. One of the common problems in peptide synthesis is contamination at some stage in the purification process through incomplete peptide chains or with the aid of products from aspect reactions. Peptide synthesizers can first-rate manipulation at once during synthesis, ensuring that pleasant-length peptides are supplied. This is achieved through the in-situ monitoring of the reaction process and using selective purification techniques to remove any contaminations during the synthesis.

Then, they can be filtered and used for mass spectrometry evaluation — which is why tracking structures are one of the fundamental characteristics that characterize top-shelf peptide synthesizers—monitoring parameters such as the absorption of UV (ultraviolet) light that indicates the successful formation of peptide bonds. These high-integrity exemplary warranty systems built into the synthesis ensure that the peptides are high-identity for purity and functionality, which is essential in drug discovery, vaccine improvement, and diagnostics programs.

Time Frame Chromatography and Analytical Tools

CM is one of the most commonly used laboratory techniques for separating, identifying, and purifying complex mixtures. This is a valuable technique in biochemical and chemical studies in which compounds seek separation based on human or chemical belongings. Chromatography takes advantage of the differences in the interaction of materials with two phases: a stationary phase (generally a solid or a liquid) and a mobile phase (typically a liquid or gas). The interplay between those levels decides at which price unique parts of the mixture travel so they can be separated.

Chromatography is crucial in purifying and quality control of synthetic compounds like peptides. Chromatography frequently cleaves the desired peptide from excess reagents, solvents, or additional by-products after the peptides are synthesized from a peptide synthesizer. This is important as high purity is required for drug discovery, diagnostics, and other biological processes with peptide programs. The properties of the most common types of chromatography that are used in peptide synthesis and evaluation, High-Performance Liquid Chromatography (HPLC) and Liquid Chromatography-Mass Spectrometry (LC-MS), are provided, with regularly used in mixture to get each separation and identification of the peptides.

One of the most common methods to purify peptides is High-Performance Liquid Chromatography (HPLC). HPLC: It separates mixtures based on the differential adsorption of compounds onto a stationary phase (usually a column packed with fine particles) while the mobile phase (a solvent or mixture of solvents) passes the column under high pressure. Retention time, or the time each factor takes to pass through the column, is the basic principle to detect and separate specific compounds from the mixture.

HPLC is particularly advantageous for peptide purification because separation may be based on hydrophobicity, length, and cost. Researchers can selectively isolate the peptides with the properties of interest by employing particular forms of columns (e.g., reverse-section columns). After that peptide purification, the peptide is collected and can be used for similar studies or advancement. HPLC is a crucial instrument in verifying peptide purity due to the highly excessive resolution of HPLC, which can separate peptides with very similar molecular weights.

Another vital use of chromatography in laboratory settings is in analyzing complicated samples. Liquid chromatography (LC) and gasoline chromatography (GC) are often used in combination with different analytical strategies in aggregation together with mass spectrometry (MS), nuclear magnetic resonance (NMR) spectroscopy, and infrared (IR) spectroscopy to provide extra precise recommendations about the structure and composition of unknowns.

Mass spectrometry has lately been commonly combined with chromatography to offer accurate data about compounds' molecular weight and structure. Once separated with chromatography, a sample is introduced into the mass spectrometer, ionized and fragmented. Reading the pattern of fragmentation tells researchers what molecular shape the compound had. This is especially true in determining complex peptides or other biomolecules that can be difficult to use chromatography independently;

Mass spectrometry can also be applied to measure the level of awareness of peptides in a pattern. This is extremely important for drugs, where the dosage of a therapeutic compound must be as accurate as possible to prove its utility through accuracy in peptide attention. In conjunction with HPLC, mass spectrometry allows researchers to confirm peptide identity, assess purity, and determine concentration in one analysis.

Another critical method for peptide and other small molecule structure elucidation, in addition to mass spectrometry, is nuclear magnetic resonance (NMR) spectroscopy. NMR employs the interactions between atomic nuclei and a magnetic discipline in a pattern placed inside it. This provides specific details regarding the chemical environment of atoms within the molecule, which enables the 3-D structure of the compound to be profiled. It has benefited NMR to study conformational adjustments in peptides and proteins, which is necessary to understand their organic function.

Another analytical technique for determining the chemical composition of peptides is infrared (IR) spectroscopy. Using a unique fingerprint that absorbs infrared light, infrared (IR) spectroscopy collects statistics about molecules' vibrations along the molecule's bonds. This technique lets us detect stimuli-responsive agencies in peptides and confirm specific chemical bonds.

Combined with chromatography, these analytical devices give researchers a complete picture of the individual peptides or compounds under assay. Chromatography helps to separate the elements of a combination, whereas analytical techniques such as mass spectrometry and carbon-13 NMR supply unique structural and compositional information. Combined, those tools develop the basis of the correct peptide identification, purification, and characterization, which are vital for utilization in drug discovery programs.

Peptide synthesizers and chromatography systems are essential laboratory equipment and tools for pharmaceutical research and development in the biochemical and molecular biology laboratory. These include complex devices that enable the efficient and correct production, purification, and characterization of peptides and other complex molecules. With the continuous enhancement of these gear alongside complementary techniques, including mass spectrometry and NMR spectroscopy, over the years and advances in this field, many extra developments in drug discovery, diagnostics, and healing improvement may also occur. When researchers know the features and programs of critical lab systems, they can ensure that their experiments are done appropriately, adequately, and at the highest standards of precision and finding.

6.2 Reagents and Chemicals

Reagents and chemicals constitute the backbone of almost all laboratory experiments. These substances are the raw materials required to perform different reactions, synthesize compounds, analyze materials, and provide the foundation for various scientific and industrial processes. Reagents matter—these reagents are very pure because even trace impurities can change experimental outcomes. In addition, chemical handling, storage, and disposal are equally important to lab fulfillment, ensuring the validity of research results, personnel, and environmental safety. Here, we learn why it is essential to select suitable reagents, the basic practices to handle exemplary reagents safely, and the proper disposal practices for suitable reagents to maintain the integrity of a laboratory.

Choosing Reagents of Better Quality

One of the most important steps when getting ready for laboratory work is to select reagents of the highest possible quality. The purity and quality of chemical reagents immediately affect the experiment's results' accuracy, reproducibility, and reliability. Laboratory reagents are available in many grades—every grade has an intended application and purity stage—and selecting the right one for a specific challenge is essential.

Understanding Reagent Grades

Reagents are generally divided between one-of-a-kind grades in keeping with cuts of purity, with the maximum commonplace grades being analytical grade, reagent grade, and USP (United States Pharmacopeia) grade. This has been adapted to satisfy the individual needs of many laboratory applications and often has to be decided depending on the accuracy expected within the experiment.

Analytical Reagent — the highest purity grade of chemicals, providing the most exact necessary results in quantitative analysis or experiments.

Chromatography, spectroscopy, and other analytical techniques use analytical reagents with high precision. These reagents are tested for impurities and should fall within strict specifications to produce accurate results.

Reagent Grade (r): Regular used chemical function (habitual laboratory techniques and chemical reactions). Although highly pure, they are not pure enough for the most sensitive experiments that provide absolute numbers. Although they can be helpful, these reagents are appropriate for experienced away synths, comedowns on assistance, and vogue commonplace chemical strategies.

USP Grade: USP grade refers to chemical compounds that meet the specifications laid down by the United States Pharmacopeia specifically for the needs of pharmaceutical programs. They are blood reagents used between the components of medicine and medical devices. USP-grade chemical compounds endure stringent trying out to fulfill the best requirements for medicinal use, ensuring that they're secure and robust while used in pharmaceutical production.

Selecting the appropriate reagent grade is crucial to ensure accurate and reproducible results. However, even trace contamination in reagents such as amino acids, resins, solvents, and coupling marketers can complicate the product or disrupt the synthesis technique, for example, in peptide synthesis. Impurities cause low yields or will infect the last product by making unwanted aspect products or failed reactions. Analytical or reagent-grade chemical compounds are regularly crucial for peptide synthesis as they maintain the excessive stage of purity, creating a tight environment required for additional experimentation or software.

Reagents purity assessment

While selecting the reagents, one of the critical factors is their purity. Residual natural solvents, outstanding natural impurities, or international chemical substances may introduce parameters that may interfere with the results of your experiments. For example, within the synthesis of peptides or different biomolecules, even hint impurities can harm the synthesis by causing through-merchandise, diminishing the yield, or changing the final structure of the peptide.

Vendors that supply laboratory chemical compounds periodically provide specifications for purities available in their products, often listing maximum allowable levels of specific contaminants. The researchers must carefully recall those specs and then select chemicals that satisfy or surpass the requirements of their particular experiment. Sometimes, rather delicate methods such as mass spectrometry (MS) or nuclear magnetic resonance (NMR) spectroscopy can be used to measure the purity of the reagents before their use in critical experiments.

For example, in peptide synthesis, the impureness of reagents (amino acids or resins) can also limit the method's performance, causing incomplete chains or requiring additional purification steps. Implementing high-purity reagents mitigates these threats and ensures the highest possible yield of the desired peptide being studied.

REAGENT STABILITY AND STORAGE

Distillation of appropriate means after having chosen the right sort of reagents out of once superb reagents, the suitable garage should be acceptable for the bare essentials. Reagents in biochemical or pharmaceutical laboratories can be especially sensitive to environmental conditions such as temperature, humidity, and light. Lack of proper garages may cause deterioration, decreased efficaciousness, or the advent of toxic byproducts.

Organic buffers and enzymes, for example, often need to be refrigerated to stabilize them. Most enzymes and biological reagents are heat-sensitive, and heating over long periods can cause them to denature or lose their activity. However, some chemicals could be hazardous or react to air and moisture. For example, reagents, including metallic sodium or potassium, are relatively reactive with water and must be stored beneath an inert environment, such as nitrogen or argon, to prevent destructive reactions.

Also, their volatility/flammability has inspired the garage of reagents. Highly flammable chemical components must be saved in heated cabinets that are a long way away from any supply of warmness and open flames. High-evaporating solvents like acetone or ethanol will be stored in sealed bins to reduce evaporation and the potential for fireplace or explosion. Labile chemistries toward light, specific vitamins, or sensitive dyes must be saved in an amber-colored bottle or dark container.

Even excellent chemical compounds can end up dangerous when mishandled, so correct labeling overdue to garage centers must also exchange the human eye sort nature. Reagents must be labeled with their name, recognition, expiration date, and danger category, ensuring that lab personnel can deal with them correctly and efficiently.

Safety Handling and Disposal

Although selecting and preserving high-quality reagents is an important consideration for any given experiment, handling and eliminating chemicals are also important aspects of laboratory safety. Many chemical compounds are hazardous and can pose sizeable hazards to human beings' fitness and the environment if they are not managed securely. Thus, lab personnel should be well educated on safe chemical dealing, the correct use of personal defensive gadgets (PPE), and excellent strategies for disposing of chemicals post-use.

PPE (Personal Protective Equipment)

Chemical Safety Personal protective devices are the first line of protection in chemical safety. Personal protective equipment (PPE) consists of all the garb and devices used to defend lab personnel from chemical exposure, splashes, and other harm. Common types of PPE include gloves, lab coats, goggles, face shields, and respirators.

Gloves- gloves are critical to chemical protection, as they save you from direct exposure to dangerous substances. There are several types of laboratory gloves: latex, nitrile, and vinyl, which provide varying degrees of protection against specific chemicals. For illustration, nitrile gloves oppose herbal solvents and many chemical substances commonly used in laboratories.

Lab Coats: Lab coats protect workers against chemical spills and splashes and protect their clothing and skin against other hazards. Lab coats must be constructed from durable, waterproof materials and sufficiently long to cover the legs and arms. In specific high-threat environments, they may be needed to deliver more defense from the ability to hearth hazards they confront.

Safety goggles and face shields: you can work with other dangerous chemical compounds in the lab, and eye safety is essential. Goggles shield against chemical splashes and dust particles, while face shields protect the face. When there is a potential risk of flying particles (from a grinder or while mixing chemical substances), face shields shall be worn like goggles.

Respirators: For laboratories that utilize unstable chemical compounds or risk inhaling poisonous fumes, respirators may be necessary. These devices protect against the inhalation of harmful gases or particulates by the respiratory system.

Safety Data Sheets for Chemicals (SDS)

A Safety Data Sheet (SDS) must accompany every chemical used within the laboratory and present important information about the chemical's properties, hazards, handling precautions, and emergency procedures. The SDS provides specific instructions and information about each reagent throughout the laboratory, and thus, it becomes an essential tool to ensure the safe utilization and garage of chemical substances. A few critical records found in an SDS are:

Substance composition and associated hazards

Measures with the first resource in case of publication or accidental ingestion.

Safe Guidelines for Coping with Garage

Spill clean up and waste casting instructions.

What are the firefighting measures in case of a chemical fire?

All chemical compounds in the laboratory should have their corresponding SDSs readily available, and lab personnel should be familiar with the laborer surroundings of each reagent, and so on. Two write-up preventive measures related to SDS safety for lab body of workers education take effects and lab personnel education to read and understand upon receiving your data.

Reagent and Chemical Disposal

Proper disposal of chemicals is part of keeping safe and maintaining environmental safety. This is important as many laboratory reagents, especially those used in synthetic chemistry or organic work, can be toxic, flammable, or otherwise hazardous to the environment. Hence, one should also follow strict guidelines for the disposal of chemicals according to local regulations.

Unless explicitly permitted, chemicals should never be put in ordinary trash or poured down the drain. Hazardous chemical compounds must be amassed in chosen waste bins and constantly labeled with the sort of waste. Drugs, almost always dissolved in an aqueous solution, should not be kept in an aquarium from the solvents used in natural synthesis. Numerous labs have the appropriate systems created by hand to eliminate particular kinds of waste properly.

Spill Response and Emergency Response

Laboratories must also have emergency and spill response handling procedures. Even a minor chemical spill poses significant threats to staff and the environment. Each lab should have emergency spill kits equipped with absorbent materials to neutralize sellers and protect agents. In addition, employees need to be educated on the proper use of those kits and the procedures to follow during a spill.

For instances of extra excessive chemical injuries such as fires or contact with extremely poisonous materials, laboratories need to have emergency showers, eyewash cubicles, and first-aid kits. Workers should know the emergency exits and evacuation routes to ensure they can leave the laboratory during a first-rate incident.

Selection, handling, and disposal of reagents and chemical compounds are prime considerations in laboratory tortures. Medical tests rely on high-quality depressants to ensure accurate and reproducible results, and the proper handling and storage methods help safeguard human safety behaviors and environmental damage. High-quality reagents, protective protocols, and a well-regarded disposal protocol would allow laboratories to maintain a secure, productive, and efficient research atmosphere. Thus, well-considered chemical substances and reagents are the cornerstone for successful scientific experimentation.

6.3 Lab Safety Protocols

It is one of the most essential aspects of laboratory safety while working in clinical, industrial, or academic settings. It consists of a range of well-connected practices intended to ensure the safety of researchers, technicians, and other staff and mitigate risks associated with the use of hazardous materials and equipment. It is essential for safety procedures to be carefully developed, enforced, and updated regularly as the capability threats that exist in laboratories, chemical, biological, or physical, exist, however, and are diverse. Whether the lab is acting on research, product trying out, or another clinical activity, following proper biosafety protocols guarantees the experimentation can progress with little demand to threaten the health and safety of lab personnel.

The risks in laboratories range from minor physical injuries, such as cuts and burns, to more severe accidents, such as chemical exposure, fire, or biological contamination. At worst, improper handling of hazardous materials or incorrect protective measures may have a lasting health impact or possibly fatal accidents. Thus, a good set of security protocols should be in place that covers everything from securing hazardous materials to emergency preparedness and first aid.

How to Treat Hazardous Materials Right

Among the most crucial aspects of laboratory protection is the appropriate management of nasty substances. Unsafe Substance refers to any chemical, organic agent, or fabric that has the potential to cause damage to human beings or the environment. These substances can be toxic, explosive, harmful, reactive, or carcinogenic and can be immediate or long-term health hazards.

Determining the hazardous nature of substances

Identification and classification of hazardous materials, based on their potential hazards, is the first step in safely managing them. Hazardous materials must be correctly identified and labeled with the appropriate hazard symbols that communicate their specific risks. They must be placed in proper bins to prevent leakages, spills, or contamination. As a rule, such a class system is controlled by direction groups like the Occupational Safety and Health Administration(OSHA) in the United States or the European Chemicals Agency(ECHA) in the European Union. It is also necessary to educate lab employees in analyzing and interpreting Safety Data Sheets (SDS) for each chemical substance, which provides specific details on the associated risks, storage, handling, and chemical disposal methods.

Personnel should be trained in risk assessment to identify the dangers of any hazardous chemical substances they are working with. In some labs, the positive chemicals (including reactive or flammable materials) may need special handling procedures, such as using explosion-proof equipment or specialized storage containers. Laboratories dealing with biological marketers retrain pathogens may additionally be required to meet guidelines for Positive guidelines for biosafety stages /g, and they inform researchers what containment, decontamination, and disposal stages are needed.

Chemical Hazards Training and Education

Practical approaches to coping in the lab are only helpful if the staff in the lab are usually accomplished thoroughly. Comprehensive protection training has to be an obligatory part of any laboratory position and ought to consist of the following areas:

Chemical Properties: Employees must be familiar with information on the chemical properties of the substances they work with. This includes knowledge about toxicity, reactivity, volatility, and flammability. Training should also reflect how these properties may influence the lab setting and personnel and what actions may be taken to decrease potential hazards.

PPE (Personal Protective Equipment): PPE such as gloves, goggles, lab coats, and face shields are essential when handling hazardous chemicals—human Efforts – Personnel should learn how to use these things and also hold them up. Gloves made from latex, for example, are unsafe when handling varied chemical substances so nitrile gloves can be recommended as a replacement. Likewise, when most unlikely dangerous organizations are being handled, respirators may be essential to stop inhalation of poisonous gases.

Proper Storage and Labeling: Responsible storage of toxic substances is one of the most critical aspects of hazardous materials management. Cabinets are safe in a place → particular chemical substances must be lethal in some safety cabinets instead → flammable garage shelves → prevents fireplace risks. Moreover, the chemicals must be stored separately in their individual containers and labelled with their names and relevant hazards.

Spill Prevention and Containment: Employees will need to be educated on how to avoid spills and leaks and what to do if one occur. Preventive measures include keeping chemicals in stable containers, using secondary containment (e.g., trays under chemical containers), and ensuring all lab equipment operates properly. In the event of a spill, personnel should know how to contain the material and follow spill response procedures properly.

Handling Dangerous and Flammable Materials

Many laboratories employ hazardous materials that present significant hazards should these materials no longer be adequately handled. These substances can be dangerous solvents, corrosive acids, or biological hazards. Some chemical substances are unstable, and working with such materials requires careful precautions to avoid breathing vapors resulting from combustion. Some examples are that volatile chemicals such as acetone or ethanol can only be used in a fume hood that helps purge toxic vapors and prevents their build-up in the lab environment. Corrosive acids, for example, sulfuric or hydrochloric acid, are handled with massive care so as not to get targeted over the pores and skin to cause a burn or the toxic vapor to inhale. In such cases, full PPE, including acid-resistant gloves and goggles, needs to be constantly put on.

This drama becomes even more difficult for laboratories that handle biologically hazardous materials. So, labs using infectious agents or genetically modified organisms must follow strict biosafety guidelines. This relies on organic protection cabinets (BSCs), which have limitations in averting the secretion of infectious materials outside the laboratory environment. Biosafety cabinets (BSCs)-------------BSCs are prepared with HEPA filters and are categorized into exclusive ranges based on the biosafety stage (BSL) of these sellers being labored with. BSL-three laboratories, for example, are intended for work on pathogens capable of creating extreme or likely deadly diseases by inhalation.

Containment and Ventilation Systems

Proper ventilation is invaluable in any lab where dangerous substances are handled: fume hoods, local exhaust airflow, and chemical fume scrubbers to attenuate airborne contaminants. For example, a fume hood pulls air from the lab area, entrapping harmful vapors or particulate count numbers and safely exhausting it outside the facility. Its importance is undeniable as it is often the first line of defense against inhalation of harmful chemicals; checking that the fume hood is operating correctly is one of the most essential parts of working in a laboratory. Lab personnel are trained to always work in fume hoods when handling hazardous materials and must frequently check that the hood operates within safety airflow ranges.

Regular Equipment Checks

A lab environment is not entirely safe, even with proper PPE, storage, and rigorous protocols, if the machinery is defective or poorly maintained. Safety system inspections (fume hoods, hearth extinguishers, eyeful stations, Protection showers, etc.) It should be checked regularly and maintained. Laboratory workers should be held accountable for reporting any problems with protection systems so that they can be repaired before they compromise laboratory safety.

Basic Emergency Procedures And First Aid

No matter where great care has been taken to restrict risk, mishaps can and do appear in laboratories. Thus, the protocols of laboratory protection should include clean, well-maintained emergency responses. They must also contain everything from chemical spills to accidents to fires and ensure that personnel members know precisely what steps to take in the event of a crisis. Keeping contingency programs in the vicinity and ensuring that every person understands the contingency plans can protect lives and assist in keeping away from severe accidents.

Chemical Spill Response

Among the lab types of injuries, chemical spills are the most frequent. The first thing we should do when facing a spill is to alert everybody who is in the lab so that they can come out or take the necessary precautions. It is essential to try to contain that spill to prevent it from spreading. This is usually done through absorbent materials such as spill pads or neutralizing agents; of course, it depends on the chemicals involved. Tackling massive spills may require emergency response teams or hazardous fabric cleanup services.

Many labs store spill reaction kits that include the essential chemicals to deal with most spills around chemicals. The kits should be easily accessible and stocked with absorbents, neutralizers, gloves, goggles, and other necessary supplies. All personnel must also recognize how to dispose of the sneaky substances after cleaning up the spill. Spills involving primarily hazardous or volatile materials may lead to same-day evacuation and professional assistance, which is essential to note.

Chapter 7: Protocols and Exercises for Peptide Synthesis.

7.1 Practical Syntheses: Getting Started

Peptide synthesis is an essential and versatile approach with significant applications in diverse fields, including biochemistry, pharmacology, molecular biology, and medicinal chemistry. Peptides, quick chains of amino acids related via peptide bonds, are essential additives in several organic systems and tactics, from enzymes and hormones to antibodies. As a result of their different organic roles, peptides have become a cornerstone in the design of therapeutic brokers, diagnostic instruments, and biomaterials. SYNTHESIZING PEPTIDES SYNTHESIS OF PEPTIDES INTEGRATION The ability to synthesize peptides with precision and high throughput is crucial for medical advancements, developing peptide-based drugs, and novel biomolecular design.

Peptide synthesis is fundamentally the assembly of amino acids in a specific order to form peptides with defined biological activities. Over the years, peptide synthesis has advanced significantly, with improvements in artificial strategies and automation rendering the provision of peptides more efficient both in small and massive quantities. Indeed, regardless of medical domains, peptide synthesis is a rare skill for laboratory experiments, drug exploration or the industry. This chapter provides an overview of experimental protocols for peptide synthesis, including critical steps, reagents, and aspects ensuring successful peptide synthesis.

Why Is Peptide Synthesis Important?

Peptides are the active ingredients of bottomless bio-technical methods. They play structural roles in proteins, function as signals, and participate in critical cellular functions such as enzyme activity, receptor interaction, and gene regulation. Others are therapeutic retailers, with uses starting from antimicrobial peptides to most cancers-targeting drugs. Many research, diagnostics, and clinical packages use synthetic peptides. In such contexts, designing and synthesizing peptides of pre-defined sequences are critical for realizing their full potential.

In drug discovery, artificial peptides are frequently used to simulate the herbal function of peptides or proteins in your natural world, granting knowledge into illness mechanisms and achievable cures. Peptides are regularly made to focus on unique receptors at the cell surface, inhibit enzyme activity, or bind to specific DNA sequences. For instance, particular peptides may be synthesized in most cancer remedies to goal tissue-precise antigens and provide cytotoxic capsules to tumor cells with a cytotoxic peptide conjugate. Analogously, peptide-based vaccines have evolved inside the subject of immunology to elicit immune responses to pathogenic or most cancer cell-specific peptides. In addition, synthetic 13mer peptides allow the researcher to investigate shape-feature relationships, as they help to understand how the composition of specific amino acids influences peptide activity, which includes interactions with other molecules.

Methods for the Synthesis of Peptides

Specific fundamental procedures are accessible for peptide combination: solid-state peptide blend (SPPS) and gaseous stage peptide blend. Each method has advantages and challenges, but mixed irregularly SPPS (SPPS) has become the method of choice for most peptide synthesis applications because of its efficiency, automation capabilities, and scalability. In SPPS, the growing peptide chain is attached to a solid support (and thus cannot freely diffuse), allowing the stepwise introduction of individual amino acids. This method allows for the simple removal of excess reagents and by-products and can be optimized for either small or large-scale peptide synthesis.

Answer-based synthesis, traditionally less normal these days, is the synthesis of peptides in remedy. This is especially useful for smaller peptides or for making a large amount of a peptide in a more reliable environment. However, scaling up is much less time-consuming and expensive than SPPS.

Knowledge of the foundations of those synthesis techniques and corresponding protocols is essential to anyone involved in peptide synthesis. Bankruptcy: In this bankruptcy, we can be aware of the realistic aspects of SPPS, as it is the most generally practiced strategy in peptide synthesis.

First Peptide Synthesis: Getting It Up And Running

The initial steps of a peptide synthesis are critical for ensuring that the process proceeds smoothly and efficiently. However, the natural synthesis process can seem daunting before being set up. With a little careful preparation and following as systematic an approach as possible, the experience can be much easier. Successful peptide synthesis depends on various variables, including the choice of resins, reagents, solvents, and an appropriate apparatus.

Setting Up Your Workstation and Equipment

Preparing the Essential Equipment, Reagents, and Workspace Before initiating peptide synthesis, preparing the essential equipment, reagents, and workspace is imperative. One of the most critical problems is choosing a suitable peptide synthesizer. Synthesizers can be broadly classified into two main types: manual (or guide) and automatic synthesizer systems. Automated solid-section peptide synthesizers are a choice for larger-scale synthesis and high throughput. This allows the inclusion of amino acids, deprotecting groups, washing the resin, and performing repetitive responsibilities of these steps to be performed by these machines. A guide synthesizer or even an unsophisticated manual manner may be employed in smaller-run syntheses.

And since that is only the synthesizer, a vast number of varied materials and methods can be necessary:

- Peptide Synthesis Apparatus: Response vessels (for guide synthesis or computerized synthesis cartridges), syringes, glassware, and frits for filtering.
- Personal Protective Equipment: Always wear appropriate personal protective equipment (PPE) when using chemicals, mainly when toxic

or irritating, as reagents (gloves, goggles, and laboratory coats will be used).

Common reagents include amino acids, coupling agents (e.g. carbodiimides or uronium salts), deprotecting agents (Example: piperidine for Fmoc protection), and solvents (e.g. dimethylformamide (DMF), dichloromethane (DCM), and acetonitrile). This requires high yield and zero facet reactions, which are both very much dependent on reagent purity.

Filtration/Drying Equipment: Filtration of your solvents and resins will ensure that you are free of contaminates that will inhibit synthesis.

Agency is a huge part of maintaining a healthy flow in your workspace. An equally essential aspect is understanding the specific role of each reagent and material in the synthesis process so they can be treated appropriately at every stage.

The Steps of Synthesizing the Understood

After setting up your workspace, learning about the list of steps involved in the synthesis process is essential. Typically, the peptide synthesis cycle includes five steps, all previously described, along with techniques to achieve them: resin loading, amino acid coupling, deprotection, washing, and cleavage. These steps ensure that the peptide is produced accurately and effectively.

Resin Loading: This stage provides the stable-segment resin to which the primary amino acid will attach. This is typically carried out by activating the amino acid's carboxyl group and allowing the amino acid's reaction with the resin-bound linker. The primary amino acid is then bonded to the resin by the linker, which provides the solid-segment assist for the subsequent peptide chain elongation.

Coupling: The subsequent amino acid is provided to the synthetic peptide chain linked to the primary amino acid. The use of a coupling reagent, including carbodiimides (e.g., EDC) or uronium salts (e.g., HATU), activates the carboxyl group of the incoming amino acid so that it can react with the free amino group of the growing peptide chain.

Deprotection: Following each coupling step, the protective group present on the amino acid at the amine (e.g., Fmoc or Boc) needs to be removed to allow for the introduction of the following amino acid. Each cycle of deprotection and coupling ensures that the peptide chain grows by one amino acid.

Washing: Between coupling and deprotection steps, the resin is thoroughly washed to remove unreacted amino acids, excess reagents, and byproducts.

It would help if you washed it properly to avoid contamination and ensure that the synthesized peptides are only part of the desired sequence.

Cleavage: Once the peptide has been fully assembled, it must be cleaved from the resin to release it. Conventional cleavage involves a strong acid, such as trifluoroacetic acid (TFA), breaking the bond between the peptide and the solid support. This is then purified by separating the compound, often consisting of separation techniques such as high-performance liquid chromatography (HPLC) and HPLC det.

Synthesis Process: How to Successfully Conduct It

Although initial experiences in peptide synthesis may seem overwhelming, success can be achieved by adhering to a few principles and avoiding common pitfalls. If each step is optimized and done carefully, you can avoid common pitfalls and get higher yields and purer peptides.

Choice of Resins and Linkers

The resin is used as the solid support of the growing peptide chain. A sound synthesis is very dependent on the correct resin being chosen. Unique resins are attractive to distinctive classes of peptides and applications. Wang resin, for instance, is typically appropriate for the synthesis of peptides that are to be cleaved as free acids, while Rink amide resin is preferred for the synthesis of peptides that are to be cleaved as amides. Standard packages utilizing maleimide (Merrifield) resins are often used for this purpose, but different specialized resins are available for peptides with distinct needs.

Also, consider the linker that binds the resin to the primary amino acid: you need to choose it cautiously. Linkers are principally chosen based on their sobriety in keeping with extraordinary response situations and the ease with which they may exist cleaved at the center of the synthesis.

Reagent Quality

Reagent quality is a critical factor in green and a hit peptide synthesis. Fresh, super reagents limit such reactions because incomplete or incorrect peptide sequences can form due to facet reactions and infection. Take sparkling piperidine to deprotect Fmoc organizations or HATU/EEDC for eco-friendly coupling reactions. Unique brother properly matched reagents per peptide mixtures.

Reaction Conditions

The peptide synthesis reaction setup needs to be optimized for each peptide. It involves manipulating parameters including, but not limited to, temperature, reaction duration, and solvent selection. That means that coupling reactions typically require heating or incubation to develop when endowed with the same kinetics of response, whereas deprotection reactions very rarely require these aggressive conditions. One should optimize for each amino acid and each synthesis to ensure the final outputs are favorable.

Washing and Solvent Choice

Washing between each step is critical to remove non-bonded reagents, excess amino acids, and reaction byproducts. If the resin is poorly washed, these impurities can be bound and interfere with further reactions. The selection of solvent is crucial to guarantee vigorous washing. Common solvents employed include DMF, DCM, and acetonitrile, but the selection is governed by the reactions' solubility and the synthesized peptide.

Monitoring Progress

The manner of synthesis is monitored regularly to pick up any troubles early. The purity and development of peptide synthesis are assessed by analytical techniques such as thin-layer chromatography (TLC) and mass spectrometry. This allows one to detect issues needing addressing, a sound reaction, nonreactivity, or any side products. An incomplete reaction or side product could be detected by analyzing the peptide at various synthesis ranges.

AbstractPeptide- synthesis is an essential process with broad applications in research, diagnostics, and therapeutic development. Understanding the fundamental concepts and techniques of peptide synthesis- including resin and reagent selection, optimizing reaction conditions, and monitoring progress- can help improve the efficiency and success of your synthetic experiments. With enjoyment, you will be more geared to face complicated peptide sequences and go beyond demanding situations that can get up. If accomplished with an interest in detail and accuracy, peptide synthesis may be a precious device in advancing technological know-how and medicine.

7.2 Exercises for Synthesising — Step By Step

The peptide synthesis is an indispensable and fundamental skill in biochemistry and molecular biology. This is particularly important for many packages, including increasing custom peptides for studies or designing peptide-primarily based therapeutics. Although the process can also sound complex, its miles are splendidly found out through a mixture of theoretical information and the arms-on exercise. In this section, a basic peptide synthesis algorithm will be broken down, guiding 1 through more advanced experiments to build knowledge in peptide synthesis.

First, we could crossover the basic peptide synthesis protocol that underlies many stable-phase peptide synthesis (SPPS) procedures. Once you become more proficient, we can explore more complex peptide synthesis challenges, such as cyclic peptides, screeded amino acids, larger peptides, or peptides containing post-translational modifications (PTMs). Each stage will provide the theory, practical action steps, and tips for success.

General Procedure for Peptide Synthesis

Due to its efficiency and scalability, solid-phase peptide synthesis (SPPS) has become the gold standard in peptide synthesis. In SPPS, residue by residue, peptides are iteratively synthesized on a solid support, where the target peptide chain is tethered to a resin support. Resin practice, amino acid loading, deprotection, coupling, wash, repeating of deprotection and coupling sequences, cleavage, and purification are the simple steps in SPPS. Here is a breakdown of the high intensity of every step in the synthesis method:

Preparation

Preparation of Required Reagents, Resins, Solvents, and Synthesis ApparatusThe immediate step before starting the solid phase synthesis of the peptide is the preparation of needed reagents, resins, solvents, and synthesis apparatus. It is essential to be precise in the instruction degree, as minor errors in preference or attention to reagents can negatively impact the response. The principal duties of the role include:

Choosing Resins and Linkers: The resin is a stationary aid for the growing peptide chain, and professional resins are selected primarily based on the peptide sequence and desired compound. Different resins, such as Wang resin, Rink amide resin, and Merrifield resin, are widely applied. The linker is a chemical entity that forms a covalent bond between the primary amino acid and the resin. Rink amide and others are examples of resins intended to release the peptide in an amidated form upon discharge. In contrast, others can help to cleave peptides at specific locations within the synthesis.

Coupling, deprotection, and activation reagents are prepared based on the peptide sequence. Common coupling reagents such as carbodiimides (EDC, DCC) or more specialized reagents such as HATU or PyBOP are often employed. Base reagents such as piperidine (for Fmoc deprotection) or acid reagents (for Boc deprotection) must also be precisely quantified.

Equipment: If limited to a synthesizer, ensure it is clear and working correctly. Although the computerized device will make many techniques less hard, you still want to realize how to use it and be capable of troubleshooting when necessary. This method requires the weighing of the resin and incorporation into the response vessel, and the entire apparatus must be purged to prevent contamination of previous batches.

Loading the Resin

This is where the first amino acid is attached to the resin. This approach involves a reaction of the N-terminal amino acid with the linker on the resin. Loading the first amino acid sets the stage for peptide chain assembly. Some essential facts to recollect at this step:

Resin Activation: The first amino acid is usually protected at its amino group to prevent unwanted reactions. The protective group on the sidechain (usually Fmoc or Boc) is gently removed right before the amino acid coupling. After the protecting group has been removed, the free amino group is then ready to be activated to undergo a coupling reaction.

Activation of the primary amino acid usually requires coupling reagents containing carbodiimides (e. g., EDC, DCC) or other reagents such as HATU or PyBOP. Apply them. These reagents initiate the carboxyl activation of the amino acid to render it more reactive toward the amino group of the resin-bound peptide.

Time and Temperature of Reaction: Coupling reaction should be carefully controlled for time and temperature. Too high temperatures may produce aspect reactions, and low reaction times may produce low/no coupling. The response under mild conditions should continue to ensure maximum efficiency.

After coupling, the first amino acid is attached to the resin, and the process can proceed.

Deprotection

Once the primary alpha-amino acid has been loaded onto the resin, the next step is to remove the protecting group from the amino potential of the amino acid. This is called deprotection. The protection set (Fmoc or Boc, amongst other things) should be cleared away to enable the utilization of the following amino acid in the pattern. There are a couple of protection organization techniques:

Fmoc protection: In an Fmoc-based peptide synthesis, the Fmoc group occupies the amino identity of the amino acid. The Fmoc group can be removed using a base such as piperidine, which cleaves the Fmoc group while leaving the remainder of the molecule intact. The following unbound amino group is now ready for coupling with the next building block.

Boc protection: In Boc-based peptide synthesis, the amino acid has the Boc group as protection, which is cleaved using an acid, e.g., TFA (trifluoroacetic acid). Next, the Boc group is removed, and the free amine is primed for coupling.

Through efficient deprotection, peptide synthesis can be halted before they occur or side reactions occur early. This requires optimized reaction time and reagent awareness to ensure whole deprotection.

Coupling

Coupling is the step through which the amino acid is added to the growing peptide chain. After removing the protective group, the amino acid in the chain is activated, and the reaction of the activated amino acid and the peptidyl-resin occurs. Here is what to work on during this step:

Importance of Coupling Efficiency: The coupling reaction's efficiency is essential for having a high yield of the desired peptide. The coupling efficiency can be optimized by introducing another amino acid, increasing the reaction time, and using high concentrations of coupling reagents such as HATU, EDC, or PyBOP.

Reaction Monitoring: Monitored after every interval step to unveil the reaction for completeness of the coupling. Thin-layer chromatography (TLC) or high-performance liquid chromatography (HPLC) analytical strategies may be used to verify incomplete coupling or side reactions. In case of an incomplete response, more coupling cycles could be performed.

Solvent selection: The solvent used for the coupling reaction has to be carefully selected based on the solubility of amino acids and reagents. Typical solvents for coupling reactions include DMF, DCM, and NMP. The solvent should not interfere with the response from here on out or cause unwanted side reactions.

Washing

Washing the resin to remove unreacted reagents and byproducts is typical after each coupling and deprotection step in the synthesis system. Washing inadequately may lead to infection or failed reactions in successive cycles. A good wash then reduces the level of side products that can introduce impurities into the final peptide.

After every response, the resin is washed with solvents (DMF, DCM, and MeOH) so that no unreacted reagents remain behind. The preference for solvent can vary based on the solubility of the reagents and the peptide being produced. In this case, a couple of washes should be performed to obliterate excess reagents.

Washing steps in fond should be very well performed; this usually involves using large volumes of solvent and allowing them long enough to dissolve and wash away unreacted reagents.

Repeat the Cycle

After the first amino acid has been successfully introduced and deprotected, and the coupling step has been completed, the deprotection–coupling–washing cycle is repeated for each amino acid in the peptide sequence. The types of cycles vary according to the length of the peptide. So, if you make a pentapeptide, you would go through the cycle 5 times to add each of the five amino acids. This process repeats until the peptide sequence is completed.

Each step of its development should be displayed, and this is vital. Analytical methods such as TLC, HPLC, or mass spectrometry will confirm that every amino acid has been delivered and that the peptide sequence is moving along correctly.

Cleavage and Purification

After synthesizing the peptide chain, this chain has to be cleaved from the resin and purified to remove any by-products or incomplete peptides. Here are the main steps of cleavage and purification:

Purification from Resin: Following synthesis, the peptide is typically cleaved from the resin using a solid acid such as Trifluoroacetic acid (TFA). The peptide is released from the resin at the cleavage point, and other protective agents at the amino acids are cleaved concurrently.

HPLC Purification: The cleaved peptide is usually purified via HSPLC. Reverse-section HPLC (RP-HPLC) is probably the most usual method because it separates peptides primarily due to their hydrophobicity. Fractions are collected and analyzed after eluting the peptides from the column using a solvent gradient.

Characterization: MS, Along with analytical HPLC or amino acid analysis. Characterization: After purifying the peptide, it should be characterized to confirm its identity and impurity.

Examples of More Complicated Peptide Syntheses

As you become more familiar with essential peptide synthesis, you may encounter more complicated peptides that involve additional steps or modifications to the basic protocol. These include cyclized peptides, peptides with non-canonical amino acids, large peptides, and peptides with publish-translational modifications (PTMs). Here is an overview of some of the more complex peptide synthesis sports:

Cyclized Peptides: Cyclization of a peptide means to form a covalent link at the two ends of the peptide, often through a disulfide bond or a ring-ending reaction. In that sense, this process requires careful control of the reaction conditions to favor the formation of the cyclic framework and avoid side reactions. Cyclized peptides are often challenging but required for specific biological functions, peptide hormone mimetics, or peptide ligands.

Peptides with Sampled Residues: Some peptides contain altered or unnatural amino acids, which can influence the synthesis process. Until then, phospho-serine or glycosylated residues will be added using specialized reagents and methods for proper incorporation. Such modifications can increase the balance of the peptide, alternate hobby, or offer new binding properties.

Significant Peptides and Proteins: Big peptides are a unique challenge, especially those greater than 50 amino acids, due to their potential for aggregation or improper folding. Design strategies such as chimeric synthesis (synthesizing the peptide in small fragments) or using multiple, tailored resins to prevent aggregation are often critical for large peptides.

PTM Peptides: Peptides that need PTMs, such as phosphorylation or acetylation, often need extra reagents to achieve selective incorporation. Phosphorylation can, for example, be introduced in vivo during synthesis through phosphorylated amino acid analogs or through enzymatic reactions.

7.3 Purification Protocols

Once you complete the peptide synthesis, purification is the next critical step to ensure the quality and usability of the final product. Both solid-phase peptide synthesis (SPPS) and solution-phase methods eventually lead to a combination of the desired peptide and other side products, including incomplete sequences (complete sequence, but particular aa missing), truncated peptides, and resin fragments, as well as byproducts from reagents used during the synthesis processing. Purification is crucial to extract such impurities and obtain a pure, homogenous peptide that can be applied for subsequent biological or chemical purposes. Especially when proving within the organic assay or clinical level, the functionality, specificity, and conventional effectiveness of your synthesized peptide with no delay depends on its purity. It will cover different purification methods, but a particular focus on HPLC is provided, including the best general recommendations for achieving good peptide purity and yield.

Why Purifying Peptides Is Important

Due to their various and significant organic functions, peptides are in high demand for drug discovery, diagnostics, immunology, and other research applications. To correctly display the organic sports anticipated for those peptides, they ought to be relatively pure so that they no longer have impurities or byproducts that would confuse experimental results or therapeutic activity. Contaminations can lift at any phase in the succession of the amalgamation, regularly because of unreacted starting materials, side items framed amid combination, or fractional deprotection. These impurities may include but are not limited to:

Truncated or Missing Peptide Sequences: The unsuccessful coupling step (i.e., due to steric hindrance) or side reaction can lead to the incomplete peptide sequence.

Side Products: Different portion reactions may additionally result in the formation of pieces or different molecules that do not have the wanted peptide sequence further than the preferred peptide.

Resin Fragments and Contaminants: Resin molecules or other reagents utilized in the synthesis path may stay attached to the peptide or contaminate the final product.

Contamination from Solvents and Reagents: Reagents and coupling marketers or deprotecting marketers can depart chemical byproducts or impurities to ship in the final peptide product.

A reliable purification protocol ensures peptides are natural enough to highlight the desired biological sports. Desiring excessive purity of the peptide could be a venture as peptides are regularly produced in small portions and feature houses that would make them hard to purify. Excessive-overall performance liquid chromatography (HPLC) is one of the most extensively used and beneficial ways of peptide purification because it splits compounds primarily based on several chemical and bodily properties.

Purification of Sample by HPLC

Amongst those, HPLC is one of the most widely used techniques for peptide purification because of its excessive determination, sensitivity, and versatility. This method separates elements of a mixture by passing the pattern via a column filled with a stationary stage at the same time as a cellular segment (solvent), which is used to elute the components at exclusive rates. Separation relies entirely on the differential interaction between the elements of the sample and the stationary phase, leading to different retention times for each compound gas chromatography technique.

HPLC Classes used in Peptide Purification

Various types of HPLCs can be used for peptide purification, each depending on the properties of the peptide and the impurities in the mixture. Other common forms of HPLC for peptide purification include reverse-phase HPLC (RP-HPLC), ion-exchange HPLC, and size-exclusion HPLC, the most commonly used types of HPLC for peptide purification.

Reverse-Phase HPLC (RP-HPLC)

Reverse-phase HPLC is the most commonly employed technique for peptide purification and is highly efficient for a wide range of peptides. Except for hydrophilic or very charged ones, it is especially handy to create peptides containing many hydrophobic or moderately hydrophobic residues. In contrast-segment HPLC, the stationary phase consists of a hydrophobic fabric, usually a C18 column, which has chemically bonded octadecylsilane (C18) agencies to silica particles. These hydrophobics are in contact with the hydrophobic regions of the peptide. At the same time, the mobile phase always consists of a combination of water and organic solvent (such as acetonitrile or methanol). As the years pass, the natural solvent awareness will increase regularly, promoting the elution of compounds according to hydrophobicity.

Peptide Hydrophobicity: Peptides containing more hydrophobic residues bind more tightly to the C18 column, while those containing more hydrophilic residues will elute earlier. It facilitates the selective elution of peptides based on their hydrophobicity via utilizing a gradient of increasing organic solvent.

Used solvent: The solvent (mobile phase) is meticulously chosen, and the gradient is optimized to elute peptides of curiosity and prevent contamination from portion products. Modifying the gradient to equilibrate the specific peptide hydrophobicity is essential for achieving high resolution and purity.

Reverse-segment HPLC is beneficial for peptides with nonpolar residues (for example, aromatic amino acids such as tryptophan, phenylalanine, and tyrosine, and aliphatic amino acids such as leucine, valine, and isoleucine) or those containing a hydrophobic motif. This methodology provides ultra chromatographic separations to isolate the peptides in >99% purity.

Ion-Exchange HPLC

Ion shot HPLC is a charming approach for purifying peptides, especially those containing charged groups or acidic or simple residues. This methodology clusters the peptides by their net cost (depending on the peptide pH and amino acid composition). Both sulfonate or amine corporations are usually included within the stationary phase of ion-alternate chromatography, and each interacts with the charged regions of the peptide. Peptides are then fractionated according to their affinity for the stationary phase—undoubtedly charged peptides bind to a cation-exchange resin Negative ubator Anonymous, and peptides bind to an anion-change resin.

Peptide charge: The charge of the peptide depends on the intracellular pH, and pKa feature values of the amino acids come into play within the peptide. In this respect, peptides with a more advantageous net charge will interact extra firmly with an anion-exchange column, while peptides with a negative charge will tightly bind to a cation-trade column.

Elution Conditions: The elution is carried out by gradually altering the pH or salt concentration of the cell segment. Those peptides displaying more favorable contacts with the desk-bound section will elute at higher salt concentrations or extra intense pH values.

The ion exchange HPLC works nicely to purify peptides that accommodate a high quantity of acidic or primary residues, including histidine, lysine, aspartic acid, or glutamic acid. This method is routinely performed in conjunction with reverse-phase HPLC as it can purify peptides based on their net charge properties and thereby allow for the resolution of peptides that may have the same hydrophobicity but different charges.

Size-Exclusion HPLC

Another method for peptide purification is size-exclusion HPLC, also known as gel filtration chromatography. With this method, we separate peptides mainly due to their molecular size. The barrier part for Desk Certain incorporates a resin that permits the smaller molecules to go into the pores and gradually make their way via the column. In contrast, giant molecules will be omitted from the pores, and the columns will tour much more quickly.

Peptide institutionalized: Heavy molecular weight peptides will be the primary to elute, followed later by smaller peptides or contaminants that may enter the pores of the stationary segment. Size-exclusion HPLC is often used with different chromatographic methods for the simultaneous purification of peptides or separating peptides for more significant impurities and protein aggregates.

Elution conditions: In general, size-exclusion chromatography is performed under isocratic conditions, with a constant mobile phase composition throughout the run, in contrast to opposite-segment and ion-alternate HPLC.

Size-exclusion HPLC is especially valuable for separating large peptides or proteins and is also helpful for purifying aggregates or contaminants that are heterogeneous in size from the desired peptide. This method is not commonly the primary method for small peptides, mainly due to lower resolution than other methods such as reverse-phase or ion-exchange chromatography.

Achieving High Purity Yield Practical Tips

To obtain a very high purity level in synthesizing and purifying peptides, each step of the method must be carefully optimized. Below are practical tips to improve the purity and maximize the peptide synthesis and purification yield.

Chapter 8: Regulatory and Ethical Considerations

Peptide production enterprise is probably one of the most dynamic and unexpectedly approaching sectors in biotechnology, utilizing peptides for many therapeutic, research, and industrial packages. Peptides, frequently short chains of amino acids, are fundamental to molecular biology and pharmaceutical research. This versatility makes them essential to new capsules, diagnostics, and remedy improvement, specifically in regions like most cancers, diabetes, infectious sicknesses, and regenerative medication. Yet, the emerging peptide marketplace raises unique regulatory, ethical, and safety considerations that must be navigated to ensure that peptide-based products are safe, effective, and compliant.

The law of peptides is a multi-layer approach involving adherence to several local and worldwide requirements, guidelines, and regulations governing the production, quality control, clinical testing, and marketing of peptide-based pharmaceuticals and therapeutics. These policies are implemented to safeguard public health and ensure that only secure and effective products make it to the market. They cover various issues, including production methods, trials, labels, and post-market surveillance. Equally significant are the ethical challenges in using peptides as they involve sensitive matters such as using peptides in clinical studies, informed consent, privacy, and the potential to misuse for non-medicinal or performance-enhancing purposes. Last but not least, peptide production's protection and environmental footprint are also of concern, especially regarding the mass manufacture of chemical wastes in laboratories and manufacturing plants to produce the peptide.

This chapter will discuss the range of regulatory frameworks that govern peptide manufacturing, the ethical issues involving peptide research and use, and the safety and environmental considerations of peptide manufacturing. By familiarizing those key areas, parties in the peptide industry—from researchers to regulatory bodies and manufacturers—will be able to ensure their work meets the highest standards of quality, safety, and ethical accountability.

8.1 Guidelines for Peptide Manufacturing

Peptide manufacturing is subjected to the law of peptide manufacturing, which is highly significant in peptides as a biologically lively subject for safety, efficacy, and consistency. Due to the great diversity of uses of peptides in therapy, biotechnology, and research, compliance with strict legislation is essential to ensure the protection of public health and the business itself. These directives control various ranges of peptide production passing guides, from the initial production of peptides to their distribution, ensuring that producers maintain high necessities in acceptable purity and caution. Without regulation, there will be a significant risk of creating poor-quality products that harm patients or nature.

Products based on peptides—therapeutics, biologics, and diagnostic equipment—are complex and often have stringent first-class guarantee system measures. Peptides are intrinsically prone to infection, degradation, and fluctuations in production, which can ultimately affect their potency and safety. In response to those concerns, international regulatory bodies, in conjunction with the US Food and Drug Administration (FDA), the European Medicines Agency (EMA), and the World Health Organization (WHO), have made consensus progress on extensive recommendations covering the peptide production method. These regulatory organizations ensure that peptide products are manufactured per established standards and are rigorously tested to confirm safety and efficacy.

Global Regulatory Standards of Importance

Global regulatory standards exist to protect consumers by ensuring that peptide-based tablets, biologics, and different therapeutic products are safe for human consumption. These guidelines outline requirements for every stage of peptide product development, from the initial studies to post-market surveillance. Regulatory bodies such as the FDA, EMA, and WHO help provide a structure for manufacturing, testing, and marketing peptide products, ensuring that they meet international satisfactory standards.

In the USA, the FDA's Center for Drug Evaluation and Research (CDER) regulates peptide-primarily based drugs and biologics. The FDA assures or maybe realizes that any peptide object slated for recuperation grade is subjected to preclinical and even scientific testing to help check its insulation, usefulness, and capacity. Before marketing a peptide-based drug, the manufacturer should file a New Drug Application (NDA) or Biologics License Application (BLA) to the FDA, including detailed data from clinical trials, information on the manufacturing process, and proposed labeling. The FDA is responsible for assessing the risk-to-benefit ratio of each peptide-based totally therapeutic product to ensure that the potential therapeutic gain exceeds the hazard posed to patients.

Similarly, the European Medicines Agency (EMA) is an essential regulatory agency regulating peptide-based medicines in all European Union (EU) member countries. As European Medicines Regulatory Network standards on approval and surveillance of medicinal products are set by the European Medicines Agency (EMA), guidelines from the EMA ensure that peptide-based drugs meet stringent quality, safety, and efficacy standards. The EMA additionally gives scientific advice and verification of scientific trials to aid developers in the regulatory method. In the EU, peptide-based products that fulfill EMA criteria can be recommended for commercialization across the 27 participant states.

Standard: WHO prequalifies medicines, including peptides, to units' global requirements for the safety, efficacy, and satisfaction (quality of manufacturing) of medicines. It analyses the drugs and the sellers of these drugs to world standards to show whether they are suited for developing countries. The WHOCNP provides that framework so that the peptides made by manufacturers worldwide, no matter where they are, can meet minimum safety and quality standards for the global trade of pharmaceutical peptides.

The other aspect of the production in the regulated requirements field is the challenge of medical trials on other peptides. Once a peptide-centered drug has been recognized, it needs to go through several stages of scientific trials to check for the protection, efficacy, and feasible facet outcomes in human beings earlier than permitted for use. These trials are meant to find optimal dosing regimens, tolerable adverse reactions, and combination effects with other drugs or conditions. GCP requires that every scientific trial be performed per Good Clinical Practices (GCP), to defend the rights and protection of subjects and to guarantee that the information amassed through a given trial has integrity.

Another important aspect of regulatory oversight is post-market surveillance. After peptide-based merchandise is permitted and made available, regulators, including the FDA, EMA, and WHO, continue to oversee their utilization to discover enamel and long-term side outcomes that have not been found during clinical trials. Such continued oversight ensures that fresh safety issues are quickly tackled and appropriate actions, including recollects or protection alerts, are put in location as required.

QC & GMP

The QC and GMP are primary to the law of peptide manufacturing. GMP recommendations are guidelines that ensure that peptide-based products are manufactured consistently and correctly. These guidelines address the protection of peptide drugs across all aspects of the manufacturing method, including raw material sourcing, packaging, and distribution, beginning from the seed. They are designed to limit risks related to contamination, variability, and errors in manufacturing peptides.

Due to the complexity of the synthesis process, particularly in the production of peptides, compliance with GMP is especially important. Peptides are usually produced through chemical or enzymatic methods, and the conditions under which they are manufactured can significantly impact their purity, potency, and overall quality. For instance, even minor variations in temperature, pH, or the focus of reactants can result in contaminants or modifications within the structure of the peptide. Therefore, tight control must be exercised at every stage of the manufacturing process to ensure that the ultimate product satisfies the required specifications.

GMP for peptide production will include some unique features, one of which is the starting materials. All pure materials used within the peptide synthesis (e.g., amino acids, reagents, solvents, and buffers) must be carefully checked and validated for purity and reproducibility. Suppose contamination or below the top quality of any of those substances occurs. In that case, it can adversely affect the ultimate product and prompt exposure to survival threats, consequent to radioactive rasping by vapor gas. Manufacturers should procure raw materials from good providers and ensure they meet a strict, pleasant standard before they are utilized in production.

Validating the manufacturing procedure is also an essential aspect of GMP. This entails checking out to ensure that the apparatus, facilities and manners used in the production of peptides are applicable to their intentions. Reactors, purification columns, and chromatography systems are all equipment that must be routinely calibrated and serviced to ensure they continue functioning at peak efficiency. Manufacturing methods should also be well documented and controlled, and every peptide product batch must undergo stringent quality testing and assessments.

This is why going via manufacturing with a suitable characterized peptide, which includes its flow, purity, efficacy, and security, is one of the most vital steps that can be performed via quality management testing. QC checking starts by examining raw materials and continuing to the finished product. Standard analytical methods applied in peptide QC include high-performance liquid chromatography (HPLC), mass spectrometry (MS), and peptide sequencing. They enable producers to confirm the composition and structure of the peptide and investigate possible impurities or contaminants in the sample. Often, the purity of peptides is analyzed by HPLC, whereas mass spectrometry can help determine the peptide's precise molecular weight and structure. On the other hand, peptide sequencing ensures that the peptide's action sequence is correct and matches the intended design.

QA systems are implemented to validate the manufacturing, instruments, and analytical methods and test the raw materials and the final product. QA systems ensure that all peptides will be produced from defined specifications, fulfilling all regulatory obligations for every batch of produced peptides. This type of verification involves reviewing manufacturing-related documentation, checking analytical techniques, and performing periodic audits to ensure that GMP requirements are followed continuously.

Supplementary to GMP, producers must ensure that their establishments are at the very least specs for cleanliness and safety. Peptide synthesis can generate risky byproducts; thus, keeping a clean and robust environment is essential to mitigate cross-contamination. As such, facilities must adhere to specific guidelines related to air quality, waste disposal, and chemical storage, and routine inspections ensure that those requirements are met.

In addition, like the physical production process, GMP includes documentation management. The production workflow must have precise and comprehensive record-keeping, from the initial peptide synthesis to the final shipment of active ingredients. This data provides an extensive record of the manufacturing process, including details on raw materials, system implementation, testing results, and any changes from standard processes. Documentation that allows traceability and accountability is crucial for recalls or investigations regarding ability protection issues.

After the approval of peptide-based capsules or merchandise and before they attain the market, compliance with GMP pointers has to persist in keeping the product excellent and protecting it for its complete shelf lifestyles. This entails the establishment of post-market surveillance programs to detect any adverse reactions or long-term side effects associated with the product. It means manufacturers must be ready to change some of their manufacturing methods if new safety concerns arise quickly or government agencies issue revised recommendations or regulations.

8.2 Ethical Divisions of Peptide Use

Ethical issues are paramount in the studies and clinical software of peptides. These issues include the common ethical dilemmas, the use of peptides in human subjects, risks and benefits, the importance of informed consent, and the security and confidentiality of information. There are also ethical issues regarding broader off-label peptide use, including for enhanced performance in athletics or in aesthetic programs.

There are also ethical issues within the recruitment of people for clinical trials. Ensuring informed consent in peptide-based studies is essential, which means that individuals who enroll in peptide-based research are fully aware of the nature of the study and potential risks and their rights as participants. This leads to the difficulty of informed consent, which ensures that participants make these voluntary, informed choices about whether or not to participate in research. Another approach for research would be for researchers to provide clear and comprehensible information about the trial, its purpose, rationale, risks and benefits, and to obtain explicit consent from participants before enrolling.

Informed Consent and Privacy

Informed consent is a fundamental normative principle in clinical studies and remedies. It is essential for subjects in peptide-related medical studies or peptide bothers to know their participation's alternative risks and merits thoroughly. Obtaining informed consent from members is part of making all records available to them in a way that is easily accessible and understandable so they are not coerced or undue influenced in their decision-making.

The other crucial ethical aspect of peptide research and treatment is the privacy and confidentiality of the individuals who took hrs and assessments. Due to the sensitivity of clinical information such as genetic and lifestyle data, physicians, hospitals, and vendors must maintain patient confidentiality throughout the clinical trial process. Personal data must be anonymized or de-identified when possible, and robust systems must be installed nearby to protect personal information from unauthorized access or misuse.

8.3 Safety and Environmental Impact

At some point along the lifecycle of peptide manufacture, the safety and environmental impact of the ensuing peptide production are vital considerations. Here, we outline the potential impacts peptide production processes can have on human health and the environment, from the synthetic origins of starting materials to the final disposal of chemical waste. Producers must avoid adverse environmental impacts and, at a minimum, ensure safe and sustainable production practices.

The Reduction of Lab Waste and Environmental Impact

The peptide synthesis process often involves using hazardous chemicals and reagents, many of which can pose risks to the environment if not handled correctly. However, in the process of peptide synthesis, waste merchandise along with solvents, residues, and byproducts are generated, and these issues need to be dealt with cautiously to minimize their environmental influence. This includes appropriate disposal, recycling, or waste disposal to reduce pollutants and environmental pollution. It advocates that manufacturers adopt green chemistry measures that concentrate on limiting the environmental impact of chemical techniques by using renewable resources and decreasing waste.

Moreover, power consumption is an essential component of the environmental effect of peptide manufacturing. If only poorly designed for energy efficiency, energy-in-depth components, including peptide synthesis and purification, could contribute to greenhouse gasoline emissions and climate trade. Manufacturers should target lower energy consumption by adopting cleaner technologies and practices, including using renewable energy sources or increasing machine efficiency to reduce energy consumption.

Chapter 9: Resources and Further Reading

Expansion of skill set and access to credible resources is necessary in any field of study or professional endeavor. Whether you are an educational researcher, a student, a practitioner, or a person with a keen interest in expanding your knowledge about a specific issue, access to the right resources can significantly impact your learning experience. This chapter provides a collection of assets to assist in additional exploration and aims to deepen your understanding of each topic. This helps introduce the topic and enables supplementary studies and career development.

This chapter provides guidance on books, journals, online platforms, databases, companies, and conferences that could serve as valuable tools in your intellectual journey. A glossary of essential terms can also be presented to ensure you are mindful of the degree's learning, which can sometimes be intricate and highly technical. These resources are advised for their educational value and practical use cases, giving a rounded support system for all interested people by expanding their understanding horizons.

This bankruptcy will replace you with the tools to dig similarly within the problem. This option is never exhaustive but provides an extensive starting line, always aiding you through the sled path of relevant materials and statistics retailers. As with anything, we are proscribed to analyse/dissect/develop an inquiry on it, and we must continue being receptive to new ideas, theories, and approaches. This chapter will help you with that process by providing pathways for your suffer-based education.

9.1 Books and Journals that are of Interest

It has a long-standing contribution to the dissemination of information with the help of books and academic journals. They offer detailed investigations of topics (historical and otherwise), introduce new findings or ideas, and provide essential critique on varying matters within a specific domain. When broadening your field of knowledge or pursuing higher levels of study, those resources serve as the foundation for building new concepts and practices in the future.

These books recommended for this bankruptcy were selected because of their skill in offering both elementary and complicated insights. Suppose you are a novice trying to understand the basics of a subject, or a professional trying to expand your knowledge. In that case, these books provide all you need — detailed summaries, factual discussions, and expert analysis. These are necessary for h/h to have a well-rounded coverage of critical subjects and are often classics in their disciplines.

For journals, peer-reviewed educational publications are essential. They are the contemporary of studies and offer robust, evidence-primarily based findings. Suppose you subscribe to or periodically review those journals. In that case, you can guarantee that you remain current with the cutting-edge developments, theories, and discussions in your region of interest. These publications offer a forum for academic debate and a platform for ideas that drive the field forward. The journals promoted in this chapter have been decided on for their high-impression contributions to the instructional community and their relevance to the problems addressed in this newsletter.

So, when you engage with these books and journals regularly, you will most likely live informed and continue to adjust your behavior to what you know about the problem. Instead, these will enhance your understanding and provide invaluable context to the information presented in this book, enabling you to forge deeper links and provide more informed analyses.

9.2 Online Resources and Databases

Information access has changed in the digital age, and the internet is the most effective tool for discovering information, networking, and developing research. The magnitude of information, from academic papers and articles to multimedia content and collaborative forums, available to users through online resources and databases is unprecedented. They are helpful for individuals looking for an in-depth study or those who want to keep themselves updated with the contemporary styles and trends in the field.

That's essentially the most significant benefit of online resources — the extendability. No more hours were spent in libraries flipping through card catalogs and dusty file cabinets. With the majority of critical academic classes, books, and statistics to be had by everyone with the aid of using the net in recent times, The online tools recommended in this chapter grant access to various of the most reputable academic databases, repositories, and research hubs, which archive a wealth of peer-reviewed articles, dissertations, books, and others. They will assist you in filtering your look consequences with the aid of using particular standards, making it more straightforward to discover applicable records swiftly and efficiently.

JSTOR, Google Scholar, and PubMed are among the most incredible instructional study databases. They grant the right of entry to articles from expert journals and periodicals, and many provide instruments for quotation administration and content material discovery. While many of those systems offer free access to some compounds, they are predominantly helpful for students and independent researchers who will not have access to paid academic subscriptions.

The primary internet resources and databases featured within this chapter provide invaluable assistance for anybody undertaking severe study or research. They help promote a better understanding of complex topics, provide a platform for collaboration, and enable easy interaction with the world's information-sharing community. Regularly checking these assets can assist you in staying atop emergent developments, accessing essential records, and interacting on the broader dialogue-taking region in your discipline of interest.

7.3 Organizations and Conferences.

Besides books, journals, and online resources, engagement with corporations and meeting intelligence are essential for professional and scholarly growth. Organizations are centers of transferring knowledge, networking, and collaboration. At the same time, conferences provide a platform to interact with thought leaders, exchange ideas, and discuss recent research findings. Both also help you meet fellow students, professionals, and mentors in your field, which builds connections that can benefit your career.

The encouraged businesses in this chapter represent not only the cornerstone organizations, advocacy organizations, and networks that help to develop the topic addressed in this book, but they frequently provide access to research, educational programs, professional certifications, and discussion forums. Many, it seems, offer opportunities to advance through networking and process listings, making it an essential resource for anyone attempting to leave their mark on the industry.

Meanwhile, conferences serve as excellent platforms for continuing education and professional growth. Besides that, they attend meetings that benefit modern-day research, discuss the demanding situations and possibilities with peers, and participate in workshops that teach new skills. Also, these events provide an excellent platform to showcase your work, receive feedback, and position yourself as a thought leader in your field. Meetings are cherished opportunities for learning, presentation, participation, and connection with the more considerable teaching or practitioner community, whether you are presenting a paper or there as a player.

You are gaining access to a wealth of information by being involved in these organizations and attending these meetings, but also placing yourself in the intellectual and professional networks that will shape and guide the future of the field.

9.4 Glossary of Key Terms

Lastly, knowing the specific vocabulary used in any field is essential for engaging with its literature and discussion. A glossary of keywords helps put you to the language of the field in mind with higher understanding and effective communication. Whether it indicates reading an academic paper, writing professional guys and females, or contributing to an investigative project, information the vocabulary makes sure that you may be in a position to navigate the concern and be counted with confidence.

The following word list provided in this bankruptcy consists of definitions of phrases that can be important to the topics and thoughts mentioned throughout the textual content. These expressions were chosen for his or her importance to the essential ideas and study areas blanketed in this e-book.

You will be able to engage with the material more deeply by getting used to those phrases, avoiding confusion, and helping you to understand more complex ideas.

The thesaurus is quite similar to the thesaurus. It is a reference device you may refer to each time you encounter odd terminology in similar studies or studies. This is a valuable resource to aid your academic or professional vocabulary development and allow you to speak and write more determinately and obviously.

Bonus Section: Exclusive Video Lessons on Peptide Synthesis and Applications

Congratulations on making it this far! To deepen your understanding of peptides and their vast applications, we've created a series of video lessons to guide you through some of the more complex concepts covered in this book. These videos will provide visual demonstrations and step-by-step explanations, bringing the science of peptides to life.

How to Access the Video Lessons

Each video lesson dives into different aspects of peptide science, from synthesis techniques to practical applications in medicine and cosmetics. Simply scan the QR code below to get started, and you'll have access to the following topics:

1. **Introduction to Peptides** – A quick overview of peptide fundamentals and their roles in biological processes.
2. **Step-by-Step Guide to Solid-Phase Peptide Synthesis (SPPS)** – Watch a practical demonstration on how to synthesize peptides using SPPS.
3. **Peptide Purification Techniques** – Learn about various purification methods, including HPLC, Gel Filtration, and Ion-Exchange Chromatography.
4. **Applications of Peptides in Skincare** – Discover how peptides are transforming the cosmetic industry.
5. **Therapeutic Uses of Peptides** – A look into peptides in medical treatments, such as cancer therapy, diabetes management, and antiviral applications.
6. **Tips for Setting Up a Peptide Synthesis Lab** – Essential lab safety protocols and equipment needs for beginners.

Ready to Start?

Scan the QR code below to access these exclusive videos. We hope these lessons help you feel confident and excited about using peptides in your own work, whether in a lab setting, skincare formulation, or therapeutic development.

Printed in Great Britain
by Amazon

60011598R00067